*Deep Learning With Python
Illustrated Guide For
Beginners And Intermediates
"Learn By Doing Approach"*

*The Future Is Here! Keras
with Tensorflow Back End*

© **Healthy Pragmatic Solutions Inc. Copyright 2018 - All rights reserved.**

The contents of this book may not be reproduced, duplicated or transmitted without direct written permission from the author.

Under no circumstances will any legal responsibility or blame be held against the publisher for any reparation, damages, or monetary loss due to the information herein, either directly or indirectly.

Legal Notice:

You cannot amend, distribute, sell, use, quote or paraphrase any part or the content within this book without the consent of the author.

Disclaimer Notice:

Please note the information contained within this document is for educational purposes only. No warranties of any kind are expressed or implied. Readers acknowledge that the author is not engaging in the rendering of legal, financial, medical or professional advice. Please consult a licensed professional before attempting any techniques outlined in this book.

By reading this document, the reader agrees that under no circumstances are is the author responsible for any losses, direct or indirect, which are incurred as a result of the use of information contained within this document, including, but not limited to, —errors, omissions, or inaccuracies.

Table of Contents

Prerequisites 8

Introduction 9

    Where to Find the Datasets? 11

Chapter 1 12

  Introduction 12

    What is deep learning? 12

    History of Deep Learning 13

    Advantages of Deep Learning 14

    Disadvantages of Deep Learning 16

    Applications of Deep Learning 16

    Conclusion 18

Chapter 2 19

  Environment Setup 19

    Downloading and Installing Anaconda 20

    Running your First Program 29

Chapter 3 39

  Theory of Artificial Neural Network 39

    How The Human Brain Works 39

    Perceptron 40

    The Activation Function 42

    Multilayer Perceptron 44

How ANN Learns? 45

Conclusion..48

Chapter 4 49

Implementing Artificial Neural Network with Keras .49

Importing Required Libraries 50

Importing the Dataset..50

Data Analysis...51

Data Preprocessing 51

One Hot Encoding the Output 51

Scaling the Data 53

Importing Keras and Subsequent Libraries 54

Adding Input and Hidden Layers.............................54

Adding the Output Layers 56

Training the Neural Network 57

Making Predictions and Evaluating the Algorithm59

Chapter 5 62

Evaluating and Tuning the ANN 62

Cross-Validation...63

Implementing Cross-Validation with Keras............64

1. Importing Required Libraries 64

2. Importing the Dataset....................................64

3. Data Analysis...64

4. Data Preprocessing 65

    5.   One Hot Encoding the Output 65

    6.   Scaling the Data 66

    Grid Search.. 69

    Implementing Grid Search with Keras 70

Chapter 6 76

  Introduction to Convolutional Neural Network......... 76

    How Computers See Images? 77

    The Convolution Operation 78

    ReLu Operation 81

    Pooling 82

    Flattening 84

    Fully Connected Layer... 85

    Conclusion.. 87

Chapter 7 88

  Image Classification with Convolutional Neural
Network 88

    Classifying Cats and Dogs Images 88

Chapter 8 105

  Introduction to Recurrent Neural Network 105

  Types of Memories in Human Brain 106

    What is an RNN? 106

    Applications of an RNN 108

    Steps of a Recurrent Neural Network................. 109

- Conclusion 111
- Chapter 9 112
  - Time Series Analysis with RNN 112
    - Downloading the Data 112
    - Data Analysis 113
    - Importing the Libraries 114
    - Loading the Dataset 115
    - Scaling the Data 115
    - Data Preprocessing 116
    - Creating RNN (LSTM) 119
    - Testing and Making Predictions 123
    - Evaluating the Algorithm 129
    - Conclusion..131
- Chapter 10 132
  - Natural Language Processing with RNN (LSTM) 132
    - Text Classification using RNN (LSTM) 133
    - LSTM Architecture 134
    - LSTM in Practice..143
    - Spam and Ham Email Classification 149
- Chapter 11 163
  - The sequence to Sequence Modeling with LSTM ...163
    - What is Sequence to Sequence Modelling? 163
    - Sequence to Sequence Architecture...................165

Creating a Chatbot Using Sequence to Sequence Model 171

Conclusion 196

What to do next? 197

Prerequisites

The book is aimed at deep learning beginners and intermediate experts. Deep learning algorithms are based on mathematical concepts such as linear algebra, calculus, and probability. Though the book targets absolute beginners in deep learning, the readers are expected to have basic knowledge of linear algebra concepts such as matrix dot products, matrix transpose, and inverse matrix etc. Similarly, readers are expected to understand multivariate calculus as well as basic concepts of probability. If you do not have these mathematical concepts, I would suggest you to first built mathematical foundations required to learn deep learning.

Introduction

Welcome to Deep Learning with Python! If you are deciding to start your career as a data scientist or machine learning expert, you have made a great decision. Deep learning is one of the hottest research areas of current times with lots of promise and great job opportunities. This book will help you build a foundation in deep learning. The book is aimed at beginners as well as intermediate deep learning experts. The book covers all the important concepts and problems that you will face as a deep learning expert. Each chapter of the book starts with theory followed by practical implementation in Python. We will solve a variety of problems in this book ranging from digit recognition to image classification and stock market prediction to NLP problems such as text classification and sentimental analysis.

The book has been divided into three main sections. The first section consists of an introduction, data pre-processing and simple densely connected neural networks. The second section consists of convolutional neural networks and its application to computer vision. We will solve the image classification problem in this section. Finally, the third section is dedicated to the recurrent neural networks where we will solve two types of problems: time series analysis and natural language processing.

The best way to take the most out of this book is to read the theory in each chapter, followed by code implementation. Do not just copy and paste the code, rather try to understand it and type it out by hand. Dataset is given for each problem; try to solve the problem with different algorithms. For instance, if in the book, a problem has been solved using the convolutional neural network; try to solve it via a recurrent neural network as well and vice versa.

Where to Find the Datasets?

The datasets for this book can be downloaded from this link:

https://www.dropbox.com/s/6r10vemdcac72av/Datasets.rar?dl=0

Chapter 1

Introduction

What is deep learning?

Deep Learning is a branch of machine learning that deals with algorithms inspired by the functionality and structures of the human brain. The term deep learning and neural networks are often used interchangeably. However, deep learning is a domain whereas neural networks are a variety of algorithms that collectively establish deep learning as a separate domain.

Deep learning algorithms are divided into three major categories: i) Densely connected neural networks or commonly known as perception, which is the most simple type of artificial neural network ii) Convolutional Neural Networks (CNN), proven to be highly efficient for image recognition tasks and iii) Recurrent Neural Networks, commonly used for natural language processing and time series prediction. All the remaining neural networks are derived from these three main categories.

Deep learning involves a layered architecture where the neural network is formed of multiple layers. The first layer of a neural network is called the input layer while the last layer is called the output layer. The input data is passed to the input layer, while the output whether continuous or discrete, is calculated from the output layer. The layers between the input and the output layers are called the hidden layers.

History of Deep Learning

ValentinGrigor'evich and Alexey Grigoryevichare credited with developing first deep learning models in 1965. They developed statistically analyzed models that used polynomial activation functions. Most optimal features from each layer were selected statistically and were handed over to the next layer. In 1970, due to unfulfilled promises of AI, the funding for the AI and DL research was substantially curtailed resulting in no significant research during that era except few individual efforts. This time period is known as the AI winter.

Significant research was undertaken in the deep learning domain from 1970 to 2010; however, no significant advantage of deep learning techniques was observed over conventional machine learning.

Though it was known that deep learning architecture, given the huge amount of data and computing power, can beat traditional machine learning algorithm, no results could prove that before 2012. The main reason was the lack of huge data sets and computing power. However, around 2012, researchers were beginning to show better results using DL algorithms. Tomas published a paper in 2012 in which he showed that word embedding technique using the DL method can improve the performance of natural language processing algorithms such as the sequence to sequence models and neural machine translation.

Currently, deep learning methods are almost always expected to beat traditional machine learning models if a huge amount of data is available.

Advantages of Deep Learning

Following are some of the advantages of deep learning:

1. Given reasonable amount machine learning models will beat traditional machine learning techniques by a huge margin. But a large amount of data is the key here. The following image explains that relationship:

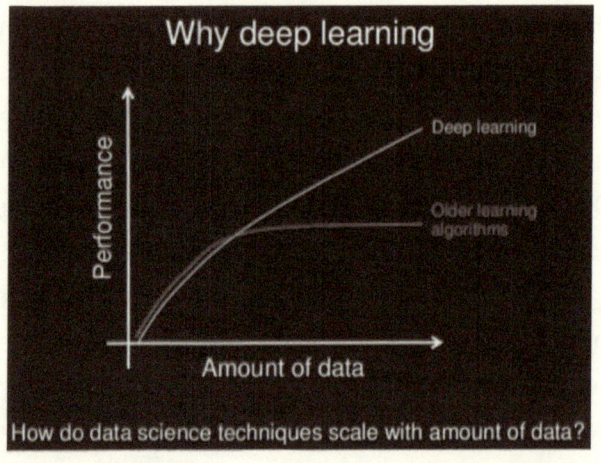

Comparison of DL and ML Performance against the amount of data (Slide from Andrew Ng)

2. DL eliminates the need for feature engineering by finding the most suitable feature for the problem.
3. DL techniques can be easily adapted to new domains such as recurrent neural networks are used to solve time series problems as well as for natural language processing.

Similarly, convolutional neural networks are also being used for NLP and computer vision.

Disadvantages of Deep Learning

1. The main disadvantage of deep learning is that it requires huge amounts of data
2. Since lots of layers and nodes are involved in a neural network, a very high-performance computing machine is required to run deep learning algorithms.
3. Deep learning algorithms work as a black box and most of the time you don't know what's happening at each node and how features are being selected.

Applications of Deep Learning

Owing to higher accuracy, deep learning is replacing conventional machine learning in almost every domain where a huge amount of data is available. Following are some of the most famous application areas of deep learning:

1. Driverless Cars

Deep learning techniques enable driverless cars to identify the obstacles on the road, increase and decrease the speed of the vehicle and take turns as per the structure of the road.

2. Gaming Industry

Deep learning is being widely employed in the gaming industry to develop intelligent characters and players that can perform as realistically as possible. [AlphaZero](), an AI created by Google using deep learning techniques recently defeated the world's best chess-playing computer. AlphaZero learned all the chess techniques in less than four hours.

3. Health Research

Research in the health sector is making use of deep learning techniques to identify different diseases. For instance, human cell data received from high power electron microscopes is fed into deep learning algorithms to identify whether the cells are cancerous or not.

4. Computer Vision

The domain of computer vision has long relied on pattern matching techniques. With the advent of machine learning, computer vision research was given a huge boost. However, the performance achieved by computer vision applications that employ deep learning has surpassed all previous research. Today deep learning is being used for image classification, video summarization, object character recognition, face recognition, and gate recognition etc.

5. Natural Language Processing

Deep learning algorithms such as neural machine translation, sequence 2 sequence models and recurrent neural network have greatly improved the performance of different natural language processing tasks such as text classification, POS tagging, sentimental analysis, answer-question system generation etc.

Conclusion

In this chapter, we introduced the concept of machine learning, followed by its brief history and pros and cons. Finally, the applications of machine learning have been discussed in detail. In the next chapter, we will see how to set up the environment required to run the scripts and codes in this book.

Chapter 2

Environment Setup

In this chapter, we will install the software that we are going to use to run our Machine learning Programs. There are several options available to implement machine learning, however, we will be using Python since most of the advanced machine learning community is working with Python for machine learning. To install Python several options available. You can simply install core Python and use a text editor like notepad to write Python programs. These programs can then be run via command line utilities. The other option is to install an Integrated Develop Environment (IDE) for Python. IDE provides a complete programming environment including Python installation, Editors and debugging tools. Most of the advanced programmers take the IDE route for Python development. We are also going to take the same route.

Anaconda is the IDE that we are going to use throughout this book. Anaconda is light, easy to install and comes with a variety of development tools. Anaconda has its own command line utility to install third party software. And the good thing is that with Anaconda, you don't have to separately install the Python environment.

Downloading and Installing Anaconda

Follow these steps to download and install anaconda. In this section, we will show the process of installing Anaconda for windows. The installation process remains almost the same for Linux and Mac.

1- Go to the following URL https://www.anaconda.com/download/
2- You will be presented with the following webpage. Select Python 3.6 version as this is currently the latest version of Python. Click the "Download" button to download the executable file. It takes 2-3 minutes to download the file depending upon the speed of your internet.

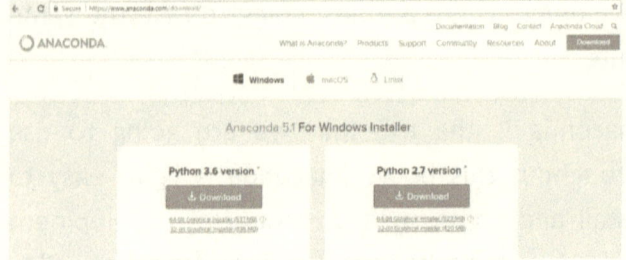

3- Once the executable file is downloaded, go to the download folder and run the executable. The name of the executable file should be similar to "Anaconda3-5.1.0-Windows-x86_64." When you run the file you will see installation wizard like the one in the following screenshot. Click "Next" button.

4- "License Agreement" dialogue box will appear. Read the license agreement and Click "I Agree" button.

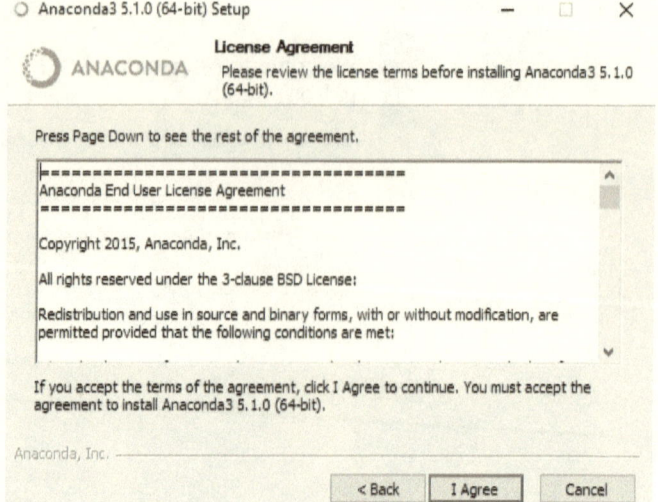

5- From the "Select Installation Type" dialogue box, check the "Just Me" radio button and click "Next" button as shown in the following screenshot.

6- Choose the installation directory (Default is preferred) from the "Choose Install Location" dialogue box and click "Next" button. You should have around 3 GB of free space in your installation directory.

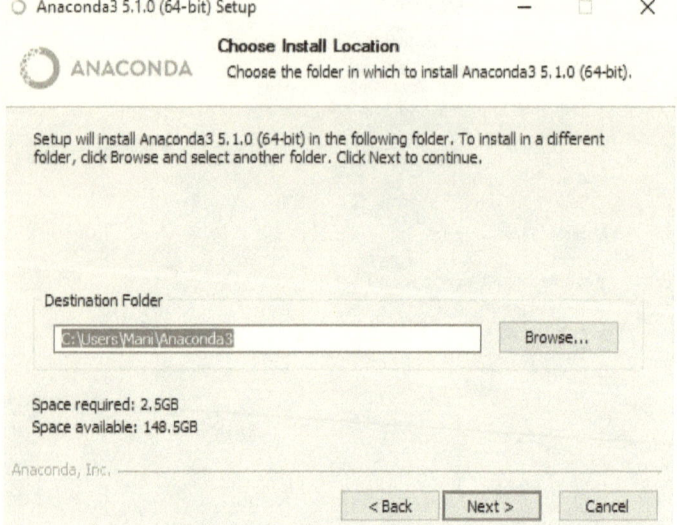

7- From the "Advanced Installation Options" dialogue box, select the second checkbox "Register Anaconda as my default Python 3.6" and click the "Install" button as shown in the following screenshot.

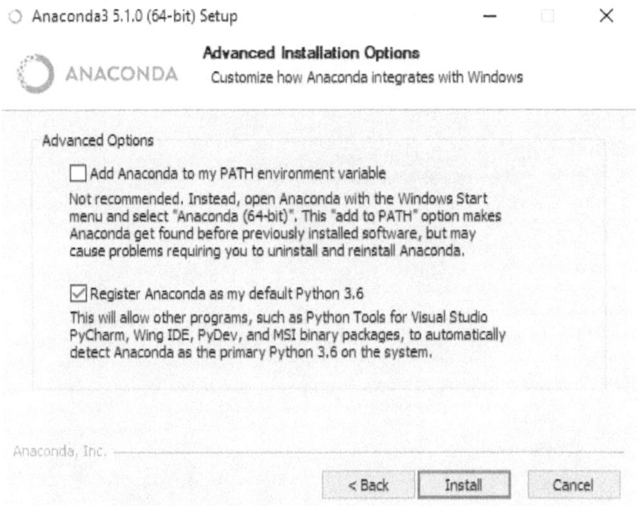

The installation process will start which can take some time to complete. Sit back and enjoy a cup of coffee.

8- Once the installation completes, click the "Next" button as shown below.

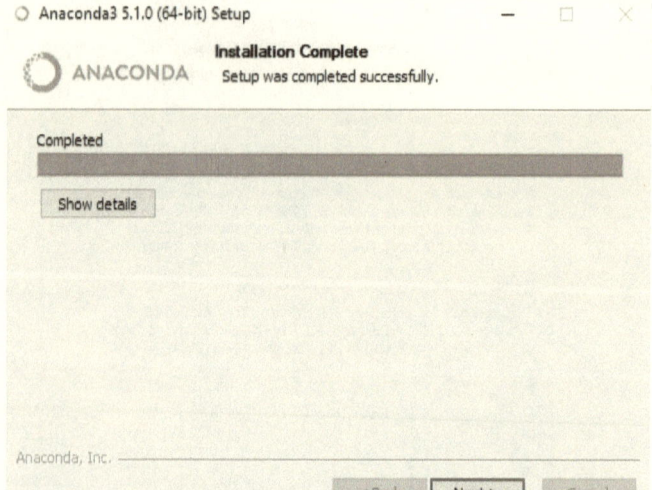

9- "Microsoft Visual Studio Code Installation" window appears, click "Skip" button.

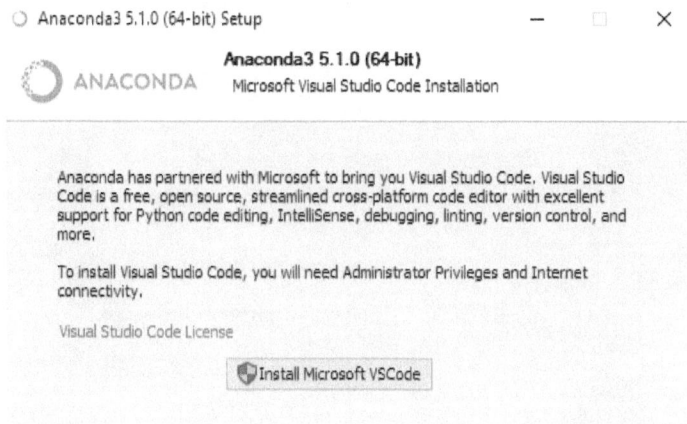

10- Congratulations, you have installed Anaconda. Uncheck both the checkboxes on the dialogue box that appears and "Finish" button.

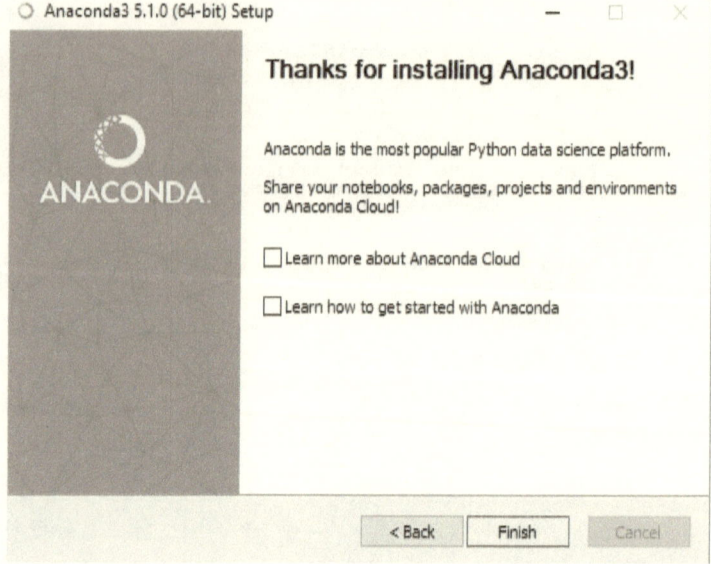

Running your First Program

We have installed environment required to run Python scripts. Now is the time to run our first program. With Anaconda, you have several ways to do so. We will see two of those in this section.

Go to your window search box and type "Anaconda Navigator" and then select the "Anaconda Navigator" application as shown below:

Anaconda Navigator
Desktop app

Folders

- **anaconda_navigator** - in site-packages
- **anaconda_navigator** - in site-packages
- **anaconda_navigator**-1.7.0-py3.6.egg-info - in site-packages
- **anaconda_navigator**-1.7.0-py3.6.egg-info - in site-packages
- **anaconda-navigator**-1.7.0-py36_0

Search suggestions

- 🔎 Anaconda Navigator - See web results >
- 🔎 anaconda navigator **youtube** >
- 🔎 anaconda navigator **windows** >
- 🔎 anaconda navigator **download** >
- 🔎 anaconda navigator **app** >

- 🔎 Anaconda Navigator

Anaconda Navigator Dashboard will appear that looks like this.

Note: It takes some time for Anaconda Navigator to start, so be patient.

From the dashboard, you can see all of the tools available to develop your Python applications. In this book, we will mostly use "Jupyter Notebook" (second from top). Though in this chapter we shall also see how to run python script via "Spyder".

Running Scripts via Jupyter Notebook

Jupyter notebook runs in your default browser. From the navigator, launch "Jupyter Notebook" (Second option from the top).

Another way to launch Jupyter is by typing "Jupyter Notebook" in the search box and selecting the "Jupyter Notebook" application from the start menu as shown below:

Jupyter Notebook
Desktop app

Folders

📄 **jupyter_notebook_**config.d - in jupyter

📄 **jupyter_notebook_**config.d - in jupyter

Documents

📄 **jupyter-notebook**-script

📄 **jupyter_notebook_**config

Search suggestions

🔍 jupyter notebook - See web results >

🔍 jupyter notebook **download** >

🔍 jupyter notebook **login** >

🔍 jupyter notebook **online** >

🔍 jupyter notebook **app** >

🔍 jupyter notebook **images** >

🔍 jupyter notebook

Jupyter notebook will launch in a new tab of your default browser.

To create a new notebook, click "new" button at the top-right corner of the Jupyter notebook dashboard. From dropdown, select "Python 3."

You will see new Python 3 notebook that looks like this:

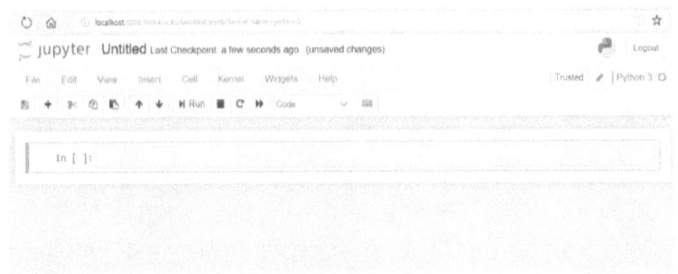

Jupyter notebook consists of cells. The python script is written inside these cells. Let's print hello world using Jupyter notebook. In the first cell of the notebook enter "print('hello world') and press

CTRL+ ENTER. The script in the first cell will be executed as shown below:

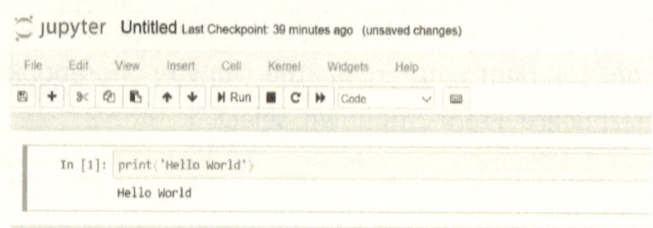

The "print" function prints the string passed to it as parameter, in the output. To create a new cell, click the "+" button from the top left menu as shown below:

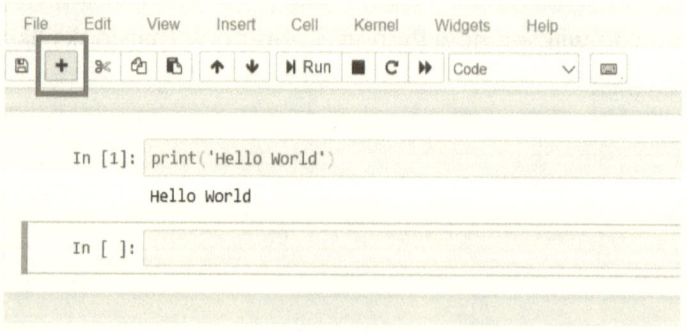

You can write Python script in the new cell and press CTRL + ENTER to execute it.

Running Scripts via Spyder

While Jupyter notebook is a good place to start writing Python programs, once you get comfortable with Python, you should move to Spyder IDE. Spyder allows us to directly create Python files. Spyder is similar to more conventional text editors with options to Run file, Run a piece of code, debug code etc.

Just like Jupyter notebook, you can run Spyder from Anaconda Navigator or directly from Start Menu. You will be presented with the following interface once you run Spyder.

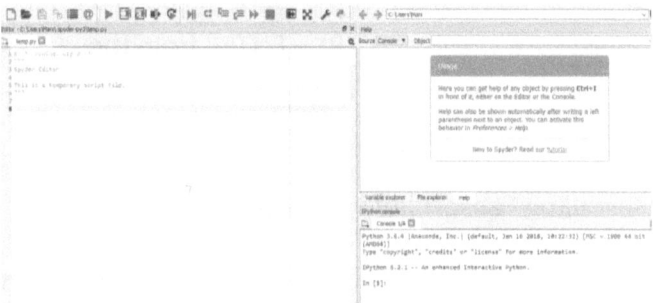

On the left side of the Spyder interface, you can see text editor; this is where you enter your script. On the bottom right you have console window. You can directly execute scripts in the console window. Furthermore, the output of the code written in the editor also appears in the console window. Let's write hello world program in Spyder.

To run the script in Python you have two options. You can either click the green triangle from the top menu or you can select the piece of code you want to execute and press CTRL + ENTER from the keyboard. You will see the output in the console window.

Installing Tensorflow and Keras

In this book, we will be using Tensorflow's Keras library for deep learning. It is very easy to install Tensorflow and Keras using anaconda prompt.

Open the anaconda command prompt by searching "Anaconda Prompt" in your search box.

Inside the command prompt, first type:

```
conda install -c conda-forge tensorflow
```

This command prompt will ask for permissions to install the packages, enter "y".

Similarly, to install Keras, the following command is used in anaconda prompt.

```
conda install -c conda-forge keras
```

What's next?

In this chapter we saw the process of setting up the environment required to run python programs. We wrote our first python program in two different editors. In the next chapter we will start our discussion about data pre-processing for machine

Chapter 3

Theory of Artificial Neural Network

If someone presents you with images of cars, bicycles, and airplanes and asks you to select the images that contain cars, you can probably do that in a matter of seconds. This is because human brains are so sophisticated that it can easily distinguish between different types of images. Two major factors play their role in human's ability to identify patterns. First is the learning; if you have never seen a car and an airplane you would never be able to tell which one is a car and which one is the airplane. The second factor is the brain's complexity which enables it to identify complex patterns.

How The Human Brain Works

The human brain consists of billions of interconnected neurons. There are three major components of a human neuron: Dendrite, Nucleus, and Axon. The dendrite is responsible for receiving messages from other neurons or sensory organs. The nucleus acts as the central processing unit and processes the information received via dendrite. Finally, the processed information is passed to Axon which transmits it to next neuron.

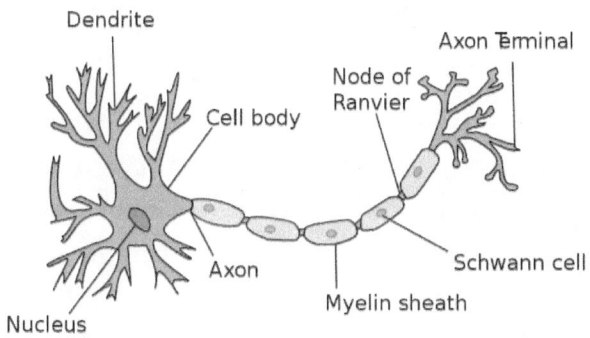

Human Neuron (Image from Wikimedia.org)

Artificial neural network (ANN) tries to imitate the human nervous system by connecting a series of neurons together to form a large network that processes the information.

Perceptron

A neuron in the human brain resembles a perceptron in ANN. Let's see how a perceptron looks like:

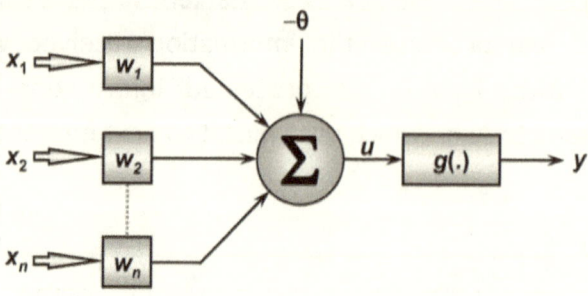

Perceptron (Image from Wikimedia.org)

A perceptron consists of one input layer and one output. The data is passed to the input layer for instance in the figure above x1, x2 to xn are the values of different feature attributes for a single data instance. The number of nodes in the input layer depends upon the number of feature attributes. Input from the input layer is multiplied with the synapse which contains some weight; each feature in the input layer is multiplied by the corresponding weight and added to the bias values. We will discuss "bias later". These values are passed to the output node where the sum of all of these multiplications is carried out using the following formula.

$$\sum_{i=1}^{n} xi.wi + bi$$

The sum is then passed through the activation function which is g(x) in the above figure which is basically the nucleus of the perceptron. The output of the activation function is treated as the final output of the perceptron. In the case of multilayer perceptron, the output of one perceptron is fed to the other perceptron and so on.

The Activation Function

In the last section, we saw the structure of perceptron. We said that the sum of inputs and weights along with bias is passed to the activation function which operates on the sum and gives the final output.

There are four types of commonly used activations functions:

1. Threshold Function

Threshold function squashes the value of the sum of dot products (multiplication of weight matrix by input matrix) of weights and inputs between 0 and 1. If the sum is less than 0, activation function outputs 0. If the sum is positive, the activation function returns 1.

2. Sigmoid Function

The sigmoid function is a smooth version of the threshold function. If the sum is greater than 0, the sigmoid function doesn't exactly return 1, it returns a value closer to 1. If the sum is less than 0, a value closer to zero is return. The sigmoid function looks like this. If the sum is zero, the value returned by the sigmoid function is 0.5.

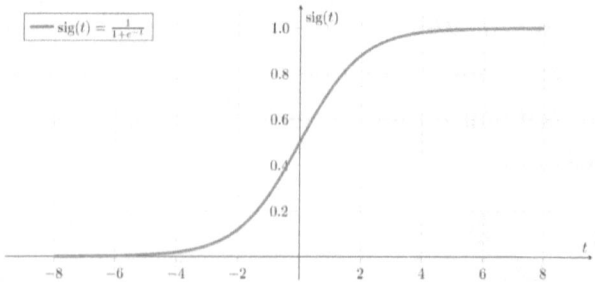

3. Rectifier

The rectifier function returns 0 if the sum of weights and input dot product is less than zero and returns the actual value if the sum is greater than zero.

4. Hyperbolic Tangent

The hyperbolic tangent function is like the sigmoid function but squashes the value between 1 and -1 and returns zero when the input is zero. Returns positive number between 1 and 0 when the input is positive and returns a negative number between 0 and -1 if the input is 0.

Multilayer Perceptron

Now we know what a perceptron is and what an activation function is. Perceptron acts like logistic regression, it can find linear boundaries. However, real-world data is not always linearly separable. In such cases, a single neuron cannot be used. A solution to find non linearly separable boundaries is to use multiple neurons and chain them in the form of layers as shown below:

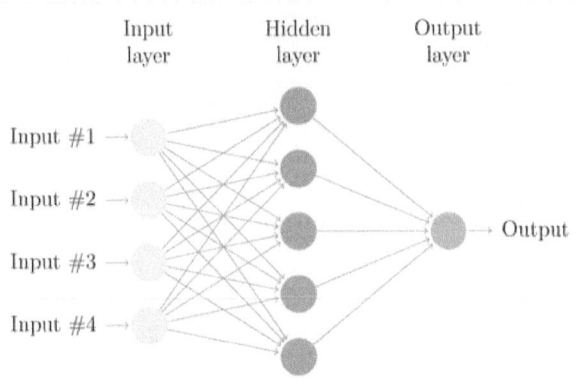

A multilayer Perceptron (Image from texample.net)

A multilayer perceptron (also known as a densely connected neural network) is the simplest neural network. It consists of an input layer which is similar to the perceptron. However, instead of directly predicting the output, it involves at least one or more hidden layers. Data from the input layer is passed to the immediate next hidden layer, which passes it to the next hidden layer and so on. Each hidden layers consists of two or more than two nodes. The dot product of inputs and weights is summed at each node and passed through an activation function. Normally, the activation function used in the hidden layers is a rectifier function while the activation function used at the output layer is a sigmoid function.

If the output is continuous or binary, one output node is enough. However, if the output involves more than two categories, more than one nodes is required at the output layer each corresponding to each category.

How ANN Learns?

Now we have seen the neural network, the next question is "how do neural networks learn". How do they know that this image is a cat or this image is a plane? Neural network learning involves two phases: Feed Forward and Back Propagation.

Feed Forward

Following are the steps involved in the feed forward phase of a neural network:

1. The input features or the values in the input layer are multiplied by the corresponding weights, bias is added to the sum and the resultant value is passed into nodes the first hidden layer.

2. Each node of the first hidden layer contains an activation function. The activation function is applied to the values received from the input layer. This process is performed at each node of the hidden layer.

3. The resultant values from each node of the first hidden layer are multiplied by the weights of the second hidden layer (if any)or the weights of the output layer if there is no second hidden layer. The result is passed to each node in the second hidden

layer where again an activation function is applied and the result value is passed to the next hidden layer. The process continues till the output layer. The values at the output layer are the final result of the algorithm.

In the beginning, the weights are randomly selected; therefore the output is usually not correct. The only thing we can control in a neural network is the weight of the synapse. The data points are immutable.

Back Propagation

1. The first step in the back propagation is to calculate the difference between the value predicted by the output layer and the actual output. This difference is called loss. The function used to calculate the loss is called loss function or cost function. Loss functions can be of many types e.g. cross entropy and mean squared error.

2. To minimize the value, we have to find the instantaneous rate of change in the value returned by the cost function with respect to each weight. Basically, we to find the values of weights for which the value

returned by the cost function is minimized. This is an optimization problem and can be solved using differentiation. For this purpose, the gradient descent algorithm can be used. Basically, in the gradient descent algorithm, the partial derivative of the cost function is found with respect to each weight. If the value is positive, it is subtracted from the current, otherwise, it is added.

This process of feed forward and back propagation continues until an acceptable amount of accuracy is achieved. One cycle involving feed forward and back propagation phase are called an epoch.

Conclusion

In this chapter, we reviewed what an ANN is without going into the mathematical details. For the sake of practical implementation, this knowledge of ANN is enough. In the next chapter, we will see how Artificial Neural Network can be implemented in Keras and how it can be used to solve prediction problems.

Chapter 4

Implementing Artificial Neural Network with Keras

In the previous chapter, we studied the theory of Artificial Neural Network. We saw how densely connected layered structure is used to learn parameters or weights that can be used to correctly predict the outcome. In this chapter, we will see how Python's Keras library can be used to implement the artificial neural network. Keras library is basically a wrapper for Tensorflow, a low-level machine learning library developed by Google. However, Tensorflow requires really deep knowledge of machine learning plus a hundred of lines of code to implement a useful neural network. However, on the other hand with Keras, you can implement an ANN with a few lines of code.

The problem that we are going to solve with our densely connected ANN is to predict the quality of wine depending upon several features.

More details about the dataset can be found at this link. We will only use the dataset for the red wine. All the datasets can be downloaded from the link mentioned in the Introduction of the book. Download the Datasets folder and place it in the D drive. The dataset used in this chapter is redwine_data.csv.

Follow these steps to Implement Artificial Neural Network with Keras.

Importing Required Libraries

Execute the following steps to import the required libraries. We will import the Keras library later.

```
import pandas as pd
importnumpy as np
importmatplotlib.pyplot as plt
%matplotlib inline
```

Importing the Dataset

The following script imports the dataset.

```
redwine_data =
pd.read_csv(r'D:\Datasets\redwine_
data.csv', sep=';')
```

The script above reads the dataset and stores it in *redwine_datadataframe*.

Data Analysis

To see how the data looks like, execute the following script. It returns the first five rows of the data.

```
redwine_data.head()
```

The output looks like this:

	fixed acidity	volatile acidity	citric acid	residual sugar	chlorides	free sulfur dioxide	total sulfur dioxide	density	pH	sulphates	alcohol	quality
0	7.4	0.70	0.00	1.9	0.076	11.0	34.0	0.9978	3.51	0.56	9.4	5
1	7.8	0.88	0.00	2.6	0.098	25.0	67.0	0.9968	3.20	0.68	9.8	5
2	7.8	0.76	0.04	2.3	0.092	15.0	54.0	0.9970	3.26	0.65	9.8	5
3	11.2	0.28	0.56	1.9	0.075	17.0	60.0	0.9980	3.16	0.58	9.8	6
4	7.4	0.70	0.00	1.9	0.076	11.0	34.0	0.9978	3.51	0.56	9.4	5

Data Preprocessing

The following script divides the data into feature and label set.

```
features=
redwine_data.iloc[:,0:11].values
labels=
redwine_data.iloc[:,11].values
```

One Hot Encoding the Output

In the output layer of the neural network, we will specify the number of output nodes. In the last chapter, we said that in case of more than two categories (binary category) we will have the N nodes in the output layer where N is the number of categories and is greater than 2. However, to use the neural network, we also have to convert our output labels array into one hot encoded matrix of K columns where K is the unique number of categories. The following script converts our label array labels into one hot encoded matrix. Execute the following script:

```
fromsklearn.preprocessing import LabelEncoder
encoder = LabelEncoder()
y = encoder.fit_transform(labels)

labels = pd.get_dummies(y).values
```

In the above script, we first use the *LabelEncoder* class from *sklearn.preprocessing* library to convert labels into a series of integers. We then used the *pandas* library's *get_dummies* method to convert this series of an integer into one hot encoded matrix.

Finally, let's split the data into training and test sets:

```
fromsklearn.model_selection import train_test_split
train_features, test_features, train_labels, test_labels = train_test_split(features, labels, test_size = 0.2, random_state = 0)
```

Scaling the Data

For Artificial Neural Network, it is necessary to scale the data, let's scale the data using standard scalar.

```
fromsklearn.preprocessing import StandardScaler
feature_scaler = StandardScaler()
train_features = feature_scaler.fit_transform(train_features)
test_features = feature_scaler.transform(test_features)
```

These are the standard steps that are required to be performed for any machine learning algorithm. Now we will see how we can use Keras library to implement Artificial Neural Network.

Importing Keras and Subsequent Libraries

We will import the Keras library and 2 of its classes: *Sequential* and *Dense*. The Sequential class is used to initialize the neural network while the dense class is used to add layers to the neural network. Execute the following script:

```
import keras
from keras.models import Sequential
from keras.layers import Dense
```

Before we can add layers to the neural network, we need to initialize the network first. Execute the following script to do so:

```
ann_clf = Sequential()
```

Adding Input and Hidden Layers

To add the input and the hidden layers, the dense class is used. The 'units' parameter of the dense class specifies the number of the output nodes in the layer. The 'kernel_initializer' parameter specifies the type of weights that should be used initially for the layer. The activation parameter specifies the activation function while the input_dim specifies the number of features in the input layers. It is important to mention that input layers is not created separately, rather the number of nodes for the input layer are specified via an input_dim parameter in the first hidden layer as shown below:

```
h1 = Dense(units = 9,
kernel_initializer = 'uniform',
activation = 'relu', input_dim =
11)
ann_clf.add(h1)
```

In the first hidden layer, we set the number of nodes to 9. As a rule of thumb, the number of nodes should be the average of input and output nodes. We have 11 input nodes and 6 output nods. The average is 8.5, hence we use 9 nodes in the hidden layers. The weights have been initialized to uniform distribution close to zero while the activation function is "relu" which is basically shorthand notation for rectifier function. In the last chapter we said that for hidden layers, relu function yields better results.

Finally, we call the add function of the Sequential class object and pass it the first hidden layer that we just created.

For the next hidden layer, everything remains same except we don't need to specify the input dimensions since the second hidden layer will calculate input dimensions from the first hidden layer. Execute the following script to add the

```
h2 = Dense(units = 9,
kernel_initializer = 'uniform',
activation = 'relu')
ann_clf.add(h2)
```

Adding the Output Layers

To create the output layer, we can use Dense class again. However, the number of units will be 6 since there are 6 categories for the output. Also, the activation function will be "softmax" since we need to predict multiple categories. For binary output, we can use the sigmoid function. Detail of Keras activation function can be found here. Execute the following script to add the output layer.

```
op = Dense(units = 6,
kernel_initializer = 'uniform',
activation = 'softmax')
ann_clf.add(op)
```

Training the Neural Network

To train the neural network we first need to call the compile method of the Sequential class object. The method takes the optimization function, the loss function and the metrics as parameters. We will use 'adam' optimization function, you can also use 'sgd' for stochastic gradient descent. Details of different types of optimizers available in Keras can be found at this link: https://keras.io/optimizers/

Since we have categorical output, we will use "categorical_crossentropy" function. If the output is binary, we need to use "binary_crossentropy" function. The details of all the loss functions available in Kerascan be found at this link: https://keras.io/losses/

The performance metrics we are going to use will be accuracy. Execute the following script to run the compile function:

```
ann_clf.compile(optimizer='adam',
loss='categorical_crossentropy',
metrics=['accuracy'])
```

Finally, to start training we need to call the fit method and pass it our training and testing set. We also need to pass the batch size and the number of epochs. For this example we only have 1599 records, therefore the batch size of 1 is good which means that weights will be updated with every observation. The number of epochs is 500. You can play around with these values to see the outcome. Execute the following script:

```
ann_clf.fit(train_features,
train_labels, batch_size=1, epochs
= 500)
```

Once you execute the above method, the neural network will start executing, after each epoch you will see the accuracy displayed as shown in the following screenshot:

```
Epoch 1/500
1279/1279 [==============================] - 1s 722us/step - loss: 1.3451 - acc: 0.4402
Epoch 2/500
1279/1279 [==============================] - 0s 301us/step - loss: 1.1223 - acc: 0.4738
Epoch 3/500
1279/1279 [==============================] - 0s 300us/step - loss: 1.0600 - acc: 0.5205
Epoch 4/500
1279/1279 [==============================] - 0s 302us/step - loss: 1.0119 - acc: 0.5601
Epoch 5/500
1279/1279 [==============================] - 0s 304us/step - loss: 0.9797 - acc: 0.5872 0s - loss: 0.9670 - acc: 0.50
Epoch 6/500
1279/1279 [==============================] - 0s 309us/step - loss: 0.9648 - acc: 0.5887
Epoch 7/500
1279/1279 [==============================] - 0s 298us/step - loss: 0.9520 - acc: 0.5864
Epoch 8/500
1279/1279 [==============================] - 0s 312us/step - loss: 0.9458 - acc: 0.5950
Epoch 9/500
1279/1279 [==============================] - 0s 309us/step - loss: 0.9412 - acc: 0.5848
Epoch 10/500
```

Making Predictions and Evaluating the Algorithm

To make predictions, we use the predict method of the Sequential object and pass it our test features as shown below:

```
predictions =
ann_clf.predict(test_features)
```

The confusion matrix and the accuracy methods used for evaluation accept test labels and predicted labels in the form of a one-dimensional array. However, we have our test labels as well as predictions in the form of one hot encoded matrix. To convert this one hot encoded matrix into one dimensional array of outputs, execute the following script:

```
predictions =
np.argmax(predictions,axis=1)
test_labels =
np.argmax(test_labels,axis=1)
```

Finally, to construct confusion matrix and calculate accuracy on the test set, we can use *confusion_matrix* and *accuracy_score* class of *sklearn.matrics* library as shown below:

```
fromsklearn.metrics import classification_report, confusion_matrix, accuracy_score
print(confusion_matrix(test_labels, predictions))
print(classification_report(test_labels, predictions))
print(accuracy_score(test_labels, predictions))
```

The output of the above script looks like this:

```
[[ 0  0  1  1  0  0]
 [ 0  0  6  5  0  0]
 [ 0  1 83 48  3  0]
 [ 0  0 33 95 14  0]
 [ 0  0  0 14 12  1]
 [ 0  0  0  1  2  0]]
             precision    recall  f1-score   support

          0       0.00      0.00      0.00         2
          1       0.00      0.00      0.00        11
          2       0.67      0.61      0.64       135
          3       0.58      0.67      0.62       142
          4       0.39      0.44      0.41        27
          5       0.00      0.00      0.00         3

avg / total       0.57      0.59      0.58       320

0.59375
```

The accuracy achieved is 59.35% in this case which is not so great. But now you know how a neural network works and what its different steps are. I would suggest you play around with different parameter such as activation functions, the optimizer and loss functions, number of nodes in the hidden layers, number of layers and so on to see if you get better results or not.

Conclusion

In this article, we saw how an artificial neural network can be implemented via Keras.We explained different steps required to implement ANN with Keras library. Enough of the basic ANN, in the next chapter we will start our discussion about a more advanced neural network famous for solving computer vision problems. Happy Coding!!!

Chapter 5

Evaluating and Tuning the ANN

In the previous chapter, we developed an Artificial Neural Network that predicted the quality of the wine. To train and test our algorithm, we divided our data into 80% training set and 20% test set. This means that we did not test our algorithm on the 80% of the data and similarly we did not train our algorithm on the 20% of the data. What if the 20% test contains information that is valuable for training? Training and testing the algorithms on a fixed subset of data may lead to variance problem.

Another problem that affects the performance comparison of different algorithms is the use of various hyperparameters such as K in the KNN algorithm and batch_size and epochs for ANN. To compare two algorithms, we need to find the parameters that result in best performance. This problem can be solved using the Grid Search algorithm.

In this chapter, we will study Cross Validation and Grid Search in detail.

Cross-Validation

Earlier we said that splitting data randomly into training and test set can lead to variance problem. Variance in performance evaluation of an algorithm refers to scenarios where an algorithm performance varies depending upon the dataset being used for training and testing.

Cross-validation is the solution to the variance problem. In cross-validation, the dataset is divided into K folds where K is any integer. Each of the K folds or partition is at least once used in the training set as well as a testing set. For instance, let's divide the dataset into 10 partitions. In the first iteration, the first 9 partitions are used for training and 10th partition is used for testing. In the second iteration, 1st to 8th and 10th partition is used for training and 9th partition is used for testing. The process continues until every partition is used at least once for testing. The final performance of the algorithm can be evaluated by taking the mean of the results from individual tests. This solves variance problem since now the result is based on the algorithm being trained and tested on the complete dataset.

Implementing Cross-Validation with Keras

In the last chapter, we used ANN to predict the quality of wine depending on several attributes. In this chapter, we will solve the same problem, however, here we will use the 10 fold cross-validation. To implement cross-validation with the Keras library, we need to follow these steps:

1. Importing Required Libraries

```
import pandas as pd
import numpy as np
importmatplotlib.pyplot as plt
%matplotlib inline
```

2. Importing the Dataset

The following script imports the dataset.

```
redwine_data =
pd.read_csv(r'D:\Datasets\redwine_data.csv', sep=';')
```

The script above reads the dataset and stores it in *redwine_datadataframe*.

3. Data Analysis

```
redwine_data.head()
```

The output looks like this:

	fixed acidity	volatile acidity	citric acid	residual sugar	chlorides	free sulfur dioxide	total sulfur dioxide	density	pH	sulphates	alcohol	quality
0	7.4	0.70	0.00	1.9	0.076	11.0	34.0	0.9978	3.51	0.56	9.4	5
1	7.8	0.88	0.00	2.6	0.098	25.0	67.0	0.9968	3.20	0.68	9.8	5
2	7.8	0.76	0.04	2.3	0.092	15.0	54.0	0.9970	3.26	0.65	9.8	5
3	11.2	0.28	0.56	1.9	0.075	17.0	60.0	0.9980	3.16	0.58	9.8	6
4	7.4	0.70	0.00	1.9	0.076	11.0	34.0	0.9978	3.51	0.56	9.4	5

4. *Data Preprocessing*

The following script divides the data into feature and label set.

```
features=
redwine_data.iloc[:,0:11].values
labels=
redwine_data.iloc[:,11].values
```

5. *One Hot Encoding the Output*

```
from sklearn.preprocessing import LabelEncoder
encoder =  LabelEncoder()
y = encoder.fit_transform(labels)
labels = pd.get_dummies(y).values
```

In the last chapter, we used *train_test_split*class of the *sklearn.model_selection* library to divide the data into training and testing sets. In the case of cross-validation, we don't need to do that, we will use the whole data set for cross-validation.

6. Scaling the Data

For Artificial Neural Network, it is necessary to scale the data, let's scale the data using a standard scalar.

```
from sklearn.preprocessing import StandardScaler
feature_scaler = StandardScaler()
features = feature_scaler.fit_transform(features)
```

These are the standard steps that are required to be performed for any machine learning algorithm. Now we will see how we can use Keras library to implement Artificial Neural Network.

7. Creating the ANN

Now let's create our artificial neural network that we will use for cross validation. We will create a method that creates the artificial neural network with two hidden layers and one output layer as we did in the last chapter. Take a look at the following script:

```python
from keras.models import Sequential
from keras.layers import Dense
deftrain_classifier():
ann_clf= Sequential()
ann_clf.add(Dense(units = 9, kernel_initializer = 'uniform', activation = 'relu', input_dim = 11))
ann_clf.add(Dense(units = 9, kernel_initializer = 'uniform', activation = 'relu'))
ann_clf.add(Dense(units = 6, kernel_initializer = 'uniform', activation = 'softmax'))
ann_clf.compile(optimizer = 'adam', loss = 'categorical_crossentropy', metrics = ['accuracy'])
returnann_clf
```

The *train_classifier* method in the above script returns the ANN.

8. Implementing Cross-Validation

To implement cross-validation in Keras, we will use *cross_val_score*class of the *sklearn.model_selection*library. However, the *cross_val_score*class accepts *sklearn* object. We have a classifier in Keras. To convert Keras classifier to sklearn compatible object, we will use *KerasClassifier* class of the *keras.wrappers.scikit_learn*library.

The KerasClassifier class accepts keras classifier and converts into the sklearn classifier. The sklearn classifier can then be used for cross_val_score function.

The classifier, feature set,and label set and the number of folds for cross-validation are passed as a parameter to the *cross_val_score* class as shown below:

```
from keras.wrappers.scikit_learn import KerasClassifier

fromsklearn.model_selection import cross_val_score


keras_ann_clf = KerasClassifier(build_fn = train_classifier, batch_size = 10, epochs = 500)

results = cross_val_score(estimator = keras_ann_clf, X = features, y = labels, cv = 10)
```

To see the accuracies returned by the *cross_val_score* class for all the five folds, you can print the list of values returned by *the cross_val_score* class as follows:

```
print(results)
```

```
[0.62500001 0.55000001 0.50625001 0.55625001 0.65        0.4875
 0.53750001 0.51250001 0.64375    0.58490567]
```

To see the average of all the accuracies, you can call mean() function on the list as shown below:

```
print(results.mean())
```

The result shows 0.5653 i.e. 56.53%.

Finally to see the standard deviation, execute the following script:

```
print(results.std())
```

The above script returns 0.0554 or 5.54% which means that tests performed on variance partitions of the dataset can return accuracy of around 51 to 61%.

Grid Search

The job of a machine learning algorithm is to find the best set of parameters or weights that yield best results. These parameters are found by the algorithms and depend upon the dataset. We cannot control these parameters.

However, there is another set of parameters that is specified before the algorithm is run. For instance value of K for the KNN algorithm, the type of kernel for the SVM algorithm, the number of estimators for the Random forest algorithm, number of nodes, batch_size, and the number of epochs for an artificial neural network and so on. These parameters can be controlled or specified.

However, we do not really know the best value for these parameters. In the last section, we set the batch_size of 10 for ANN with 500 epochs. However, we do not know if this is the ideal. What if the algorithm performs better with a batch size of 5 and 300 epochs?

Grid Search algorithm helps us solve this problem. Basically what Grid Search algorithm does is, it automatically finds best parameters for a particular algorithm from a set of parameters.

Implementing Grid Search with Keras

To implement grid search with keras, the first 1 to 4 and 6,7 steps are the same steps that we performed in the cross-validation. We don't need one hot encoding for grid search. Therefore, update the step 5 i.e. script for one hot encoding step as follows:

```
from sklearn.preprocessing import LabelEncoder
encoder =  LabelEncoder()
labels = encoder.fit_transform(labels)
```

After executing the script described in those seven steps, executing the following script:

```
from keras.wrappers.scikit_learn import KerasClassifier
keras_ann_clf = KerasClassifier(build_fn = train_classifier)
```

The above script converts keras classifier into sklearn classifier. We will use *GridSearchCV* class of the *sklearn.model_selection* library; therefore we need to convert keras classifier into sklearn classifier as we did in the cross-validation section:

Creating Parameter Dictionary

Grid search algorithm doesn't just randomly run and finds all the best parameters for an algorithm because it can take years. For instance, if the grid search algorithm starts testing each value for the batch_size parameter of ANN from 1 to 500, the algorithm has to run a minimum of 500 times and that's just for 1 parameter. Therefore a set of values for each parameter that you want to test is passed to the GridSearchCV class. These set of parameters and are passed in the form of a dictionary.

Suppose we want to test different values for batch_size, and a number of epochs for the ANN, we can create a dictionary that looks like this:

```
grid_params = {'batch_size': [10, 5],
               'epochs': [500, 300]
               }
```

In the script above we create a dictionary named, "grid_params" the keys for the dictionary are the names of the parameters and the values of the dictionary correspond to values of the parameter. From these set of values, the Grid Search algorithm will return the best combination.

It is important to mention here that the grid search algorithm can take a lot of time depending upon the values that you want to test and the number of folds for the cross-validation.

For instance, in our case, we have 2 values for the batch_size parameter and 2 values for epochs. Total possible combinations in this case are 2 x 2 = 4. Multiply this value with the number of folds e.g. 10. That makes it 40 executions. Out of 40, 20 executions have 500 epochs, while 20 executions have 300 epochs, which makes a total of 16 thousand epochs. This can slow down the algorithm a bit.

Creating Parameter Dictionary

To execute the Grid Search, we need to create object of GridSearchCV class and pass it the classifier, the parameter dictionary that we just created, the performance evaluation metrics (we will use accuracy), and cross validation folds. Execute the following script to create GridSearch object.

```
fromsklearn.model_selection import GridSearchCV
gs= GridSearchCV(estimator = keras_ann_clf,
```

```
param_grid = grid_params,
scoring = 'accuracy',
cv = 10)
```

The final step is to call fit method on the GridSearchCV object and pass it the training and test set as shown below:

```
gs_result = gs.fit(features, labels)
```

This can take a bit of time to execute.

Once the above script executes, the last and final step is to see the parameters selected by grid search, to do so you can use the best_params_ attribute of the GridSearchCV class as shown below:

```
print(gs_result.best_params_)
```

The output looks like this:

```
{'batch_size': 10, 'epochs': 300}
```

The output shows that the best parameters from the given set of parameters are batch_size of 10 and 300 epochs.

Best performance is achieved with the aforementioned parameters values. Finally to see the accuracy achieved using the most optimal parameters, execute the following script:

```
print(gs_result.best_score_)
```

0.5941213258286429

You can see that with a batch size of 10 and epochs 300, we received average accuracy of 59.41% which is greater than 56.53% we obtained in the last section with a batch size of 10 and 500 epochs.

Conclusion

In this chapter, we saw how we can use cross-validation to cope with the variance problem. We also studied how grid search can be used to select the best parameters for an algorithm. In the next chapter, we will see the Convolutional Neural Network (CNN) which is an extremely power algorithm and is commonly used for computer vision problems.

Chapter 6

Introduction to Convolutional Neural Network

In the last chapter, we implemented a densely connected neural network where each node in the previous layer was connected to each node in the next layer. This is the simplest architecture of a neural network. However many advanced neural networks have also been developed for specialized tasks. One such neural network is Convolutional Neural Network (CNN), which is used to identify patterns in the data that is in the form of 2-D maps, such as images.

CNN is commonly used for computer vision problems as such image classification, video summarization etc. In this chapter, we will study the intuition behind the convolutional neural network. In the next chapter, we will implement the convolutional neural network in Python to solve image classification problem.

How Computers See Images?

It is important to mention here that computers treat images differently than humans. We, as humans see lines and curves and different colors in an image. While the computers can only see numbers. In case of a grayscale image, an image is treated as a two-dimensional array of values between 0 and 255 where 0 is full black while 255 is full white. In the case of colored images, there are 3 channels of values between 0 and 255, one each for red, green and blue color.

Suppose, we have the following grayscale image, it may look smile to us as seen on the left side, but to computers, it is nothing more than a two-dimensional array of zeros and ones.

Note: For the sake of simplicity, here 1 indicates black pixel while 0 indicates white pixel. Actually, 1 is for white and 0 is for black.

 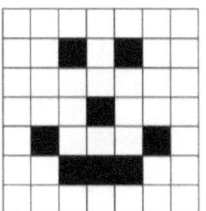 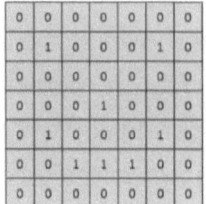

Now we know what an image looks like in a computer's memory. Let's start our discussion about CNN.

Following are the steps required to be performed for the execution of a convolutional neural network.

1. The Convolution Operation
2. ReLu
3. Pooling
4. Flattening
5. Fully connected layer

The Convolution Operation

The first step involved in a convolutional neural network is the convolution operation.

In convolution operation, we have an input image, as shown at the left side of the following image and a feature detector or also known as convolution operator or kernel, as shown in the form of a 3 x 3 matrix in the following image.

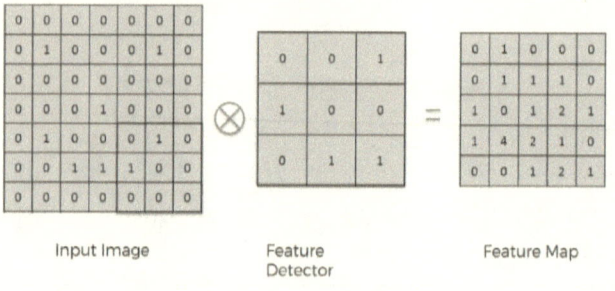

Fig 6.1: Convolution Operation

In the convolution operation, the feature detector is moved over the image from left to right and top to bottom. The corresponding pixels are multiplied and the products of all corresponding pixels are added together. It is important to mention the concept of a stride here. A stride is a step taken by the feature detector. A stride of 1 means, feature detector is moved one step towards right or bottom.

For instance, at the bottom right corner of the input image, you can see the following matrix of numbers:

0	1	0
1	0	0
0	0	0

When you multiply each pixel value of the above matrix with the corresponding pixel values of feature detector and add the products together you get the following value:

0x0 + 1x0 + 0x1 + 1x1 + 0x0 + 0x0 + 0x0 + 0x1 + 0x1 = 1

You can see 1 in the bottom right corner of the Feature map.

In reality, there are multiple feature detectors resulting in multiple feature maps as shown in the following image. An image has multiple features and one feature detector can detect one feature only. Therefore it makes sense to have multiple feature detectors.

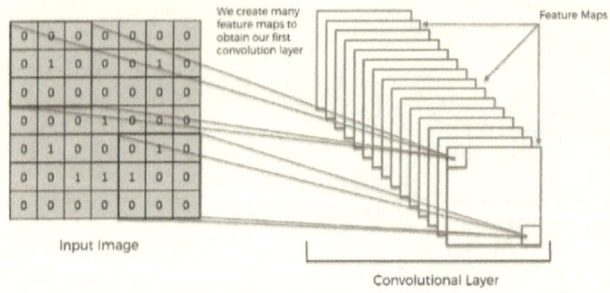

Fig 6.2: Feature Detectors

ReLu Operation

Once the feature maps have been generated as a result of convolution operation, the next step is to perform ReLu operation. We studied Relu Operation in detail in Chapter 3. Convolution operation is linear. However, real-life images are not linear. To introduce non-linearity in the image, the ReLu operation is performed.

In the ReLu, all the negative values in the feature maps are replaced by zero. The positive values are not changed.

Suppose we have the following feature map:

-4	2	1	-2
1	-1	8	0
3	-3	1	4
1	0	1	-2

When the Relu function is applied on the feature map, the resultant feature map looks like this:

0	2	1	0
1	-0	8	0
3	0	1	4
1	0	1	0

Pooling

Before looking at pooling, let us understand spatial invariance. Take a look at the following images of a cheetah. The first image is the actual image, in the second image the orientation is changed while in the third, the image is distorted. However, in all the three cases, we are able to Identity that this is a cheetah. We can identify, nose, ears, and eyes even if they are not in the same position in the image.

Fig 6.3: Spatial Invariance

This is called spatial invariance where irrespective of the location of the feature, you can identify it. Pooling helps achieve spatial invariance.

Like convolution, in pooling, we have a filter of any size. Suppose we have a filter of size 2x2. We place it on the image and the maximum value from that stride is selected. This is called max pooling. In some cases sum, average and min pooling is also used. Take a look at the following image:

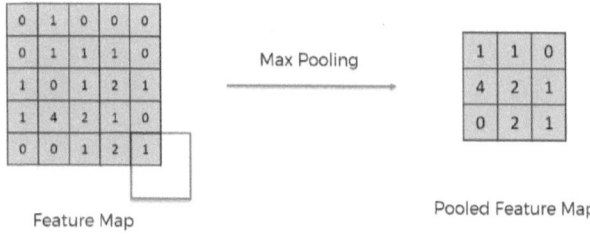

Fig 6.4: Max Pooling

In the 2nd and 3rd row and 3rd and 4th columns, we have values 1, 1, 1 and 2. The maximum value is 2; you can see that maximum value in the pooled feature map on the right.

Where there are no values to compare as is the case with the bottom right corner where we only have 1. It is forwarded to the pooled feature map.

Flattening

To detect more features from an image, the output of the convolutional neural network i.e. pooled feature maps are used as input to the densely connected neural networks. Pooled feature maps are in the form of 2-D arrays. However, the input to the densely connected neural network should be in the form of a feature vector or 1-D array. In the flattening step, pooled feature maps are flattened as shown in the following figure:

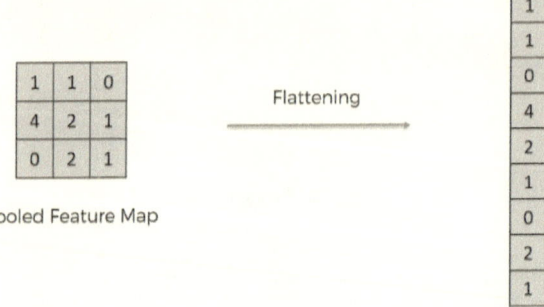

Fig 6.5: Flattening

In case of multiple pooled feature maps, they can be stacked together to form one long column vector.

Fully Connected Layer

The last and final step of a convolutional neural network is to feed the flattened pooled feature maps into a densely connected neural network which we studied in chapter 4 and 5 as shown in Fig 6.6. This step is performed to find more features from the existing set of features generated by convolution, relu and pooling steps.

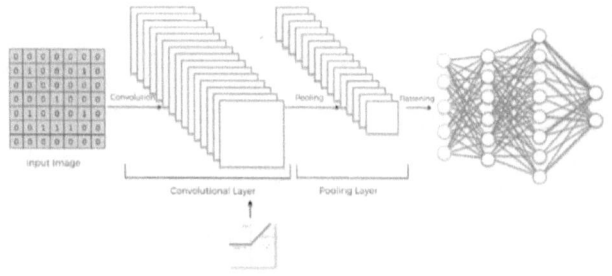

Fig 6.6: Feature Detectors

To visualize CNN, check this tool developed by Adam Harley. This tool can be used to visualize CNN. You can write the digit in the digit box at the top left corner, and the CNN will detect what you have written. The CNN used by this tool has 2 layers of convolution, relu and pooling operations, followed by two densely connected layers as shown in the image below:

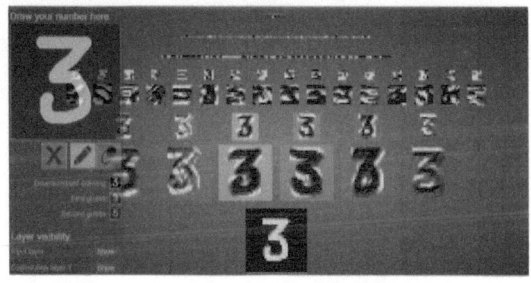

Fig 6.7: Visualizing CNN (Check this tool)

Note: The pooling and ReLu operations have been combined together.

I would suggest you play around with this tool and build more intuition about the convolutional neural network.

Conclusion

In this chapter, we studied the theory behind CNN. We saw all the steps involved in the execution of a CNN. In the next chapter, we will implement CNN in Python and solve an image classification problem.

Chapter 7

Image Classification with Convolutional Neural Network

In the previous chapter, we studied what a Convolutional Neural Network (CNN) is and what are the various steps involved in the development of the convolutional neural network.

In this chapter, we will move forward with the implementation of the convolutional neural network in Python. We know that a convolutional neural network can learn to identify the related features on a 2D map such as images. In this chapter, we will solve the image classification task with CNN. Given a set of images, the task is to predict whether an image contains a cat or a dog.

Classifying Cats and Dogs Images

This is a very famous problem and the dataset for this problem is also available on google datasets. The original dataset contains around 25000, images. However, for the sake of simplicity and ease of download, we will use only 10,000 images for training and testing. The dataset for this chapter can be found in the animal_datafolder which is available inside the Datasets folder that can be downloaded from the link described in the description section of this book.

Understanding the Data

Before we can move forward with coding the CNN, we need to understand how the dataset is arranged since we do not have a single CSV file that can be used as a data source. Rather we have a set of images. To implement machine learning models, we need dependent and independent variables. The independent variables are used to predict the dependent variables.

To convert a set of images into independent and dependent variables, we have several options. One of the options is to name each image as cats or dogs, depending upon the image type, followed by a unique id. The images can then be used as independent variables. For creating a vector of dependent variables, we can parse each image name and depending upon the value found i.e. cat or dog we can add a certain entry into the dependent variable vector. This is cumbersome and involves some custom code.

However, there is a better solution inKeras. We just need to arrange images in the certain folder structure and Keras will automatically detect the dependent and independent variables. Furthermore, Keras will also detect the training and test sets depending upon the directory structures.

Therefore, for Keras to understand the independent and dependent variable we have to arrange images in a certain folder structure.

First, you need a top-level folder that contains both the training and test set. Inside the top-level folder, you need two folders, one each for training and test sets. Inside both the training and test sets, you need folders for each category. For instance, if you have two types of images i.e. cats and dogs, you will need to create folders for cats and dogs inside the training and test set folders.

The directory structure we will use in this Chapter looks like this:

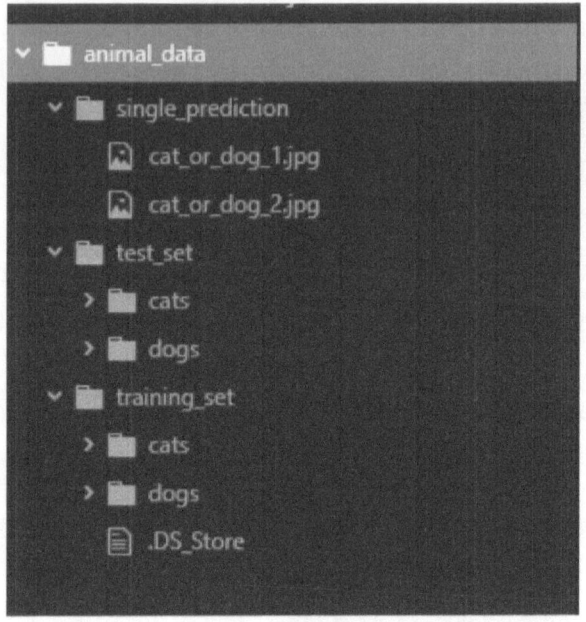

You can see that we have one top-level folder animal_data which contains test_set and training_set folders. The animal_data folder also contains single_predictionfolders, which we will not use in this Chapter. Both the test_set and the training_set folder contains dogs and cats folder. We will use 8 thousand images for training and 2 thousand for testing. Therefore, the training_set folder contains 8 thousand images. Both the cats and dogs folder inside the training_set folder contain 4 thousand images each. Similarly, cats and dogs folders in the test_set folders contain 1 thousand images each. If you open the cats folder, you will see images of the cats, while the dog's folder contains images of the dogs.

The good thing about this arrangement of the dataset is that you can use it for any problem. For instance, if you want to classify images of cars and trucks, simply create folders for cars and trucks in both training_set and test_set folder and you are good to go.

We have prepared the data for Keras. Now is the time to code the convolutional neural networks in Keras. Follow these steps:

1. Importing Libraries

As always, the first step is to import the required libraries. Execute the following script:

```
from keras.models import Sequential
from keras.layers import Conv2D
from keras.layers import MaxPooling2D
from keras.layers import Flatten
from keras.layers import Dense
```

Let's see what each class does. The *Sequential* class from the *keras.models* library is used to build a neural network that contains multiple layers. Remember, we used this in the ANN section as well. The next class is the *Conv2D* class from the *keras.layers* library. As the name suggests, this layer is used to perform the convolution and relu function on the input image. *MaxPooling2D* class from *keras.layers* library is used to apply max pooling to the feature map generated as a result of convolution operation. Next, the *Flatten* class is used to flatten the pooled feature map into a column vector. Finally, the *Dense* class is used to create a densely connected layer.

Initialize the Convolutional Neural Network

The first step in building any neural network is to initialize it. We did this in the ANN section as well. This step is the same for almost all the neural networks:

```
cnn_clf = Sequential()
```

We have initialized our neural network. The next step is to add layers to this network.

Adding Convolution & Pooling Layers

The first layer that we will add to our convolutional neural network is the convolution layer. To create a convolution layer, we will create an object of Conv2D class.

The first parameter to the Conv2D class is the number of feature detectors, or filters or kernels. It is important to mention that the number of feature maps generated because of convolution layer will be equal to the number of feature detectors. We will use 32 feature detectors for the dogs and cats image classification. You can play around with more.

The second parameter to the Conv2D class is the dimensions of the feature detector. We know that the feature detector is a matrix that we slide over the image to generate feature maps. We need to specify the rows and column of feature detector inside the parenthesis. We will use a 3 x 3 feature detector.

The third parameter is the input shape parameter. This is basically the shape of the input image that the convolutional neural network expects.

The images that we are going to use are colored images with different dimensions. In the last chapter, we discussed that black and white images have only one channel. On the other hand, colored images have three channels: red, green and blue. Though our images have different dimensions we will preprocess to have dimensions of 64 x 64 pixels.

So for our input_shape parameter, we will specify (64, 64, 3); 64 x 64 for the pixels and three for the RGB channels. Finally, we have to specify the activation function. As discussed in the last chapter, the activation function we will use is the ReLu function.

Execute the following script to create first convolution layer.

```
conv1 = Conv2D (32, (3, 3),
input_shape = (64, 64, 3),
activation = 'relu')
```

To add this convolution layer to our neural network, we will use the *add* function as follows:

```
cnn_clf.add(conv1)
```

The next step is to add pooling layer. For that purpose, we will use the MaxPooling2D class. We only need to pass the dimensions of the pooling filter. We will use a 2 x 2 pooling filter for this task.

```
pool1 = MaxPooling2D(pool_size =
(2, 2))
```

Finally, let's add the pooling layer to our CNN.

```
cnn_clf.add(pool1 )
```

Let's add one more convolution and pooling layer and add it to our convolutional neural network.

To create and add a second convolutional layer, execute the following script:

```
conv2 = Conv2D(32, (3, 3),
activation = 'relu')
cnn_clf.add(conv2)
```

It is important to mention that for the second convolutional layer, we do not need to specify the input shape parameter since the second convolutional can automatically detect the input_shape from the previous layer.

To create and add second pooling layer, execute the following script:

```
pool2 = MaxPooling2D(pool_size =
(2, 2))
cnn_clf.add(pool2)
```

Flattening the Input

We need to flatten the input data before we can use it with the densely connected neural network. To flatten input data from the previous convolution and pooling layers, we can use Flatten class as follows:

```
cnn_clf.add(Flatten())
```

Adding Dense Layers

The final step in a convolutional neural network is to input the data from flatten layer to the densely connected layer. The densely connected layer is the same layer that we used to create ANN in chapter 3 and 4. We will use one hidden layer of 128 nodes. The following script adds a hidden layer:

```
h1 = Dense(units = 128, activation = 'relu')
cnn_clf.add(h1)
```

In the output layer, we will use *sigmoid* activation function since we have binary outcome. Execute the following script:

```
h2 = Dense(units = 1, activation = 'sigmoid')
cnn_clf.add(h2)
```

Compiling the CNN

The compilation step of CNN is similar to ANN. We use the *compile* function and pass it the optimizer function, loss function and the metrics as shown below:

```
cnn_clf.compile(optimizer =
'adam', loss =
'binary_crossentropy', metrics =
['accuracy'])
```

Image Augmentation

In our dataset, we have 8 thousand images to train our model on. This is actually a very small number and can result in over-fitting. Over-fitting refers to the phenomena where a machine learning models results in higher accuracy on training data but predicts test instances with lower accuracy.

In simple machine learning problems, we only have to find a relation between the dependents and the independent variables. However, in the case of images, the machine learning model has to find relations between different parts of the image.

Therefore, for image classification, a huge number of images is required.

Luckily, Keras contains a class named image data generator that can help generate images of different types from the existing images so that we can have an increased number of images in the training set.

ImageDataGenerator class generates copies of images and then performs tasks like switching image orientation, shrinks its size, distorting the image to create new unique images from the existing images.

Look at the following script to see the working of ImageDataGenerator class.

```
fromkeras.preprocessing.image
import ImageDataGenerator


train_setgen =
ImageDataGenerator(rescale =
1./255,
shear_range = 0.2,
zoom_range = 0.2,
horizontal_flip = True)
```

In the script above, we first import the *ImageDataGenerator* from the keras.preprocessing.image library. We then create an object of the class. In the constructor, we need to pass the values for rescaling, shear_range, zoom_range and horizontal_flip. These are all the different types of transformations that will be applied to the images in order to different types of training data. In the script above, we apply these transformations for the training set.

For the test set, we want to keep everything as it is except that we will scale the images. Execute the following script to create an image data generator for the test class.

```
test_setgen =
ImageDataGenerator(rescale =
1./255)
```

Creating Training and Test Sets

We have created the classed for transforming training and testing set. We will use these classes to create training and tests sets that we are going to use. Take a look at the following script:

```
training_data =
train_setgen.flow_from_directory(r
```

```
'D:/Datasets/animal_data/training_set',
target_size = (64, 64),
batch_size = 32,
class_mode = 'binary')

test_data = train_setgen.flow_from_directory('D:/Datasets/animal_data/test_set',
target_size = (64, 64),
batch_size = 32,
class_mode = 'binary')
```

Since, our images are stored in a local directory; therefore, we need to call *flow_from_directory* method in order to apply transformation. The method takes four parameters. The first parameter is the path of the directory that contains the images that you want to transform.

The second parameter is the *target_size*parameter that specifies the dimensions of the images. In the first convolutional layer, we specified that our images are going to be 64 x 64 pixels. Therefore, pass a tuple of (64, 64) to the *target_size*parameter.

The second parameter is the batch size parameter, which specifies the batch size of the images that you want to generate. Finally, since we have two classes in the output i.e. cats and dogs, therefore, we set *class_mode* parameter as a binary.

Finding Accuracy

We have training data and testing data in the desired format. The next step is to train the model and find the accuracy. Execute the following script to do so:

```
cnn_clf.fit_generator(training_data,
steps_per_epoch = (8000/32),
epochs = 25,
validation_data = test_data,
validation_steps = (2000/32))
```

To train our model we need to call the *fit_generator* method on the *cnn_clf* object. We need to pass the training data as the first parameter. The second parameter is the steps_per_epochs which is basically batch size. Since we have 32 images in each batch, we divide the number of images in a batch by the total number of batches so that each image is used once for training during the batch. The number of epochs is the number of times you want to repeat the training overall batch. The validation_data parameter received test_data as a parameter. Finally, for the test set, you again have to pass the number of batches. Here again, we have 2000 batches and in one batch 32 images are tested. Therefore, two test all the 2000 images in the test set we divide 2000 by 32.

The above script will run for 25 epochs. You will see improve in both accuracy and validation accuracy with each epoch. The accuracy is the accuracy achieve on the training set while the validation accuracy is the accuracy achieved on validation or test. Once the algorithm is trained, the result set looks like this:

```
Epoch 16/25
250/250 [==============================] - 90s 301ms/step - loss: 0.3400 - acc: 0.8467 - val_loss: 0.4624 - val_acc: 0.8070
Epoch 17/25
250/250 [==============================] - 103s 412ms/step - loss: 0.3257 - acc: 0.8600 - val_loss: 0.4533 - val_acc: 0.8040
Epoch 18/25
250/250 [==============================] - 90s 394ms/step - loss: 0.3187 - acc: 0.8596 - val_loss: 0.4771 - val_acc: 0.7895
Epoch 19/25
250/250 [==============================] - 90s 306ms/step - loss: 0.2998 - acc: 0.8715 - val_loss: 0.4685 - val_acc: 0.7935
Epoch 20/25
250/250 [==============================] - 98s 392ms/step - loss: 0.2851 - acc: 0.8822 - val_loss: 0.4720 - val_acc: 0.7900
Epoch 21/25
250/250 [==============================] - 97s 388ms/step - loss: 0.2767 - acc: 0.8856 - val_loss: 0.5011 - val_acc: 0.7850
Epoch 22/25
250/250 [==============================] - 97s 387ms/step - loss: 0.2537 - acc: 0.8901 - val_loss: 0.4685 - val_acc: 0.8050
Epoch 23/25
250/250 [==============================] - 97s 390ms/step - loss: 0.2563 - acc: 0.8950 - val_loss: 0.5136 - val_acc: 0.7950
Epoch 24/25
250/250 [==============================] - 97s 388ms/step - loss: 0.2350 - acc: 0.9007 - val_loss: 0.5401 - val_acc: 0.7855
Epoch 25/25
250/250 [==============================] - 106s 425ms/step - loss: 0.2290 - acc: 0.9040 - val_loss: 0.5235 - val_acc: 0.8010
```

You can see from the output that after 25 epochs, we receive an accuracy of 90.40% and validation accuracy of 80.10%, which is not bad given the fact that we used very simple parameters. You can try to improve accuracy by decreasing the batch size and increasing the number of epochs. Another way to increase the accuracy of the algorithm is to increase the number of convolutional layers. Finally, if you are unable to increase accuracy despite performing these steps, increase the target size of the image by adding more pixels along the horizontal and vertical axis. This will add more information to the algorithm, which may result in improved accuracy.

Conclusion

In this Chapter, we implemented a Convolutional Neural Network in Keras to classify images. We saw the different steps required for image classification. From the next chapter, we will start our discussion about Recurrent Neural Network (RNN). We will see different types of RNN and will try to solve problems such as time series analysis and text classification etc.

Chapter 8

Introduction to Recurrent Neural Network

In the previous chapter, we concluded our discussion about Convolutional Neural Network (CNN), which is an advanced type of Artificial Neural Network. CNN is commonly used for computer vision problems. In this chapter, we will study Recurrent Neural Network which is another advanced type of artificial neural network. and is commonly used to solve time series analysis problems. Recently a specialized form of RNN, known as Long Short-Term Memory (LSTM) is widely being used for time series analysis and natural language processing tasks such as machine translation and sequence to sequence modeling.

In this chapter, we will briefly review the theory behind RNN. In the next few chapters, we will see the application of RNN for time series analysis and natural language processing.

Types of Memories in Human Brain

We know that Artificial Neural Networks were modeled to imitate the human brain. The human brain has two types of memories: Long-Term Memory and Short Term Memory. For instance, with help of long-term memory, we can recognize the movie that we saw five years ago. We can recognize the faces of our close friends even if we see them after years. On the other hand, short-term memory normally helps us remember a thing for a short span of time and then it fades. For instance, we do not remember what we had in food three weeks ago. In the human brain, temporal lobe stores long-term memories and the frontal lobe is responsible for short-term memory.

What is an RNN?

An Artificial Neural Network is trained using repeated cycles of feed-forward and back-propagation steps. In feed-forward step, the error is calculated and in back-propagation, the weights are updated to reduce the error. Once the error is minimized, the weights can be saved. These weights constitute long-term memory.

Recurrent Neural Networks differ from other neural networks in a way that in case of RNN, the output of the neuron is fed into itself as an input. This replicates the behavior of short-term memory of the human brain. Figure 8.1 shows the architecture of an RNN.

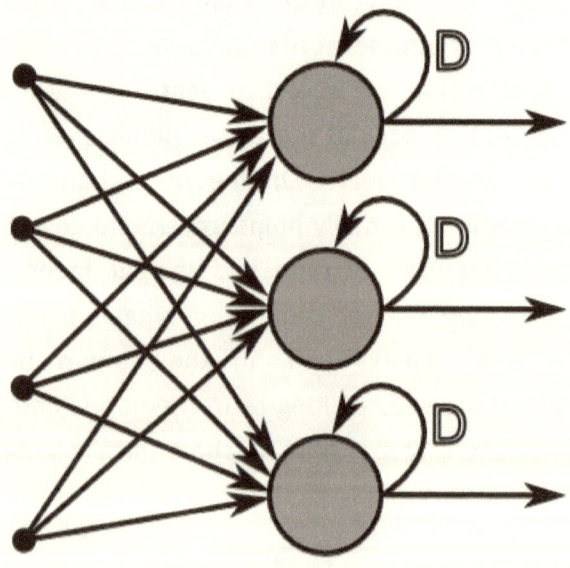

Figure 8.1: Architecture of a Recurrent Neural Network

Applications of an RNN

The applications of RNN can be divided into three major categories:

- **One to Many**

These are the type of applications where there is one input and multiple outputs. An example of such an application can be the image descriptor where an image is an input to the system and the description of the image is the output.

- **Many To One**

Text classification is a typical example of such type of applications. For instance, in a spam classification application, a text string containing multiple words is fed as input to the system while the output is the label ham or spam.

- **Many to Many**

Machine translation is one of the finest examples of the applications of RNN belonging to Many to Many category.

For instance, if you have a sentence "The boy eats an apple and he likes mangoes too". If you change the above sentence to "The girl eats an apple", a leaves the rest of the sentence as it is, a RNN application will give a warning that "he" in "he likes mangoes" should be changed to "she" since RNN keeps tracks of previous information and it found the word "girl there" instead of a boy.

Steps of a Recurrent Neural Network

As we saw earlier, a recurrent neural network is very similar to an Artificial Neural Network. The only addition is the feedback data from the hidden layer into itself. Like an ANN, the RNN also consists of multiple cycles of feed forward and back propagation steps.

Feedforward

Following are feed forward steps for the recurrent neural network:

1. For each node in the hidden layers, the input values are multiplied with the corresponding weight matrix of that particular hidden layer, which can be represented as Wxh * X. Here Wxh is the weight matrix for the hidden layer h and X is the input vector. The asterisk here refers to the dot product between the two vectors.

2. Since at each node of the RNN, we input the data from the previous hidden state to the new hidden state. This can be represented as Whh * h[t-1]. Here Whh is the weight matrix for hidden state h and h[t-1] is the output of the hidden state at the previous timestamp i.e. t-1.

3. The output from step 2 and 3 is added together along with the bias and is used as the input for the current hidden state.

4. The activation function is applied to the input at the hidden state which generates new values.

These steps are performed at each node in the network until the output nodes are reached.

Backpropagation

The backpropagation steps for the RNN are exactly the same as an ordinary ANN:

1. The error between the actual output and calculated output is calculated using a loss function which is normally negative log

likelihood since the activation function used at the output node is usually softmax.
2. The error is back propagated throughout the network and the weights are updated accordingly.

Conclusion

In this chapter, we saw the basic principle behind the recurrent neural network, which is the feedback mechanism from the output of the hidden state from the previous timestamp to the new timestamp. This allows RNN to remember what happened previously which in is helpful for predicting sequential data. In the next chapter, we will see how LSTM, which is a version of RNN, solves time series analysis problems such as future stock price prediction.

Chapter 9

Time Series Analysis with RNN

In the previous chapter, we started our discussion about RNN. We said that RNN can be used for predicting sequential data. In this chapter, we will predict future stock prices for Microsoft with the help of LSTM which is one of the most commonly used versions of the LSTM. Let's get straight to the task:

Downloading the Data

The stock market data of Microsoft can be downloaded from a variety of sources. However, for the sake of this article, we downloaded the data from the following link:

https://finance.yahoo.com/quote/MSFT/history/?guccounter=1

If you go the above link, you will see the latest stock data for Microsoft. You can apply a date filter to filter for any desired time period.

In this chapter, to train our RNN, we will use the Microsoft stock data for 5years from 1-Jan-2012 to 31-Jan-2016. To test our algorithm and make evaluations, we will use the Microsoft data for one month i.e. January 2017.

For the sake of this chapter, the training data has been downloaded and saved in the Datasets folder of the D drive by the name ms-training.csv. Similarly, the test file has been saved as ms-testing.csv in the same folder.

Data Analysis

If you open the file, you will see that it contains seven columns: Date, Open, High, Low, Close, Adj Close and Volume. We are only interested in predicting the future stock price at the opening of the day.

Figure 9.1 shows how the price at the opening of the day faired over a period of five years. The pattern is extremely irregular and with a normal ANN, it is very hard to capture such a pattern. This is where LSTM comes to play. The LSTM is capable of finding such a sequential pattern in the data.

We have visualized the data, now is the time to do some real programming. We will follow the traditional machine learning pipeline for this purpose.

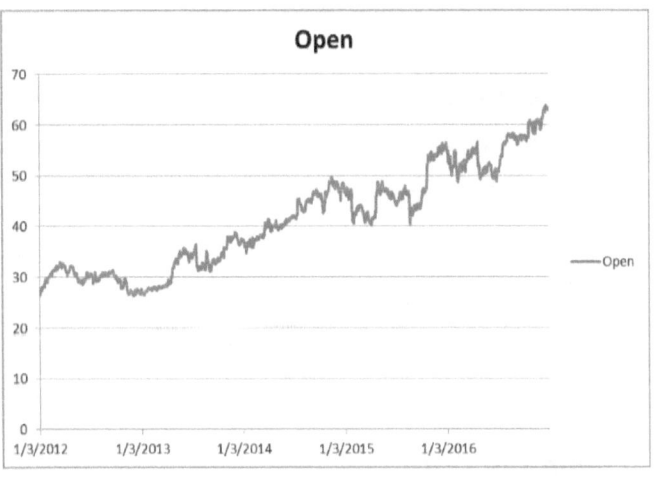

Figure 9.1: Variation of Opening Stock Price over the years

Importing the Libraries

```
Import numpy as np
```

```
importmatplotlib.pyplot as plt
import pandas as pd
```

Loading the Dataset

The following script loads the dataset and stores it in the stock_datadataframe.

```
stock_data =
pd.read_csv('D:\Datasets\ms-
training.csv')
```

We do not the whole dataset for training. We only need the values in the "Open" column. The following script extracts values from the open column and stores it in a numpy array:

```
stock_training=
stock_data.iloc[:, 1:2].values
```

Scaling the Data

The input data to an LSTM should always be scaled. For the sake of this article, we will normalize our data so that it is scaled. Execute the following script to do so:

```
fromsklearn.preprocessing import
MinMaxScaler
normalizer =
MinMaxScaler(feature_range = (0,
1))
stock_training_scaled =
normalizer.fit_transform(stock_tra
ining)
```

In the script above we use the *MinMaxScaler* class from *sklearn.preprocessing* library. We use the *feature_range* parameter to scale the data between 0 and 1.

Data Preprocessing

Data pre-processing is one of the most important steps in a recurrent neural network. We need to input the data to an RNN in such a way that it contains the information about previous time steps as well as the current time steps since we said earlier that an RNN learns from previous data.

To train our algorithm, we will use the opening stock price over the previous 60 days. The number 60 has been chosen because various researchers have tried to predict the opening stock price at any given day based on previous 10, 20, and 50 and 60 days and the results show that the best performance is achieved on using last 60 days as input.

In other words, we need to include past 60-time steps in the training data. The output or the label will contain only single value i.e. the opening stock price at 61st day.

The following script creates training and test sets:

```
train_features = []
train_labels = []
for i in range(60, 1258):
train_features.append(stock_traini
ng_scaled[i-60:i, 0])
train_labels.append(stock_training
_scaled[i, 0])
```

In the script above, we declare two lists: train_features and train_labels. We then loop through 61st (index starts from zero in numpy array, hence the 60th index contains 61st element) to the last element in the stock_training_scaled array.

In each loop cycle, we append the previous 60 values to the train_features list and the current value to the train_labels list. In the output, we add the current element.

Finally, we need to convert our training and test set to numpy arrays. The following script does that:

```
train_features, train_labels =
np.array(train_features),
np.array(train_labels)
```

Finally, we need to reshape our dataset into a specific format. The LSTM expects data in a three-dimensional format where the first dimension is the number of records in the training dataset, the second dimension is the number of time steps while the third dimension is the number of indicators. The number of indicators simply refers to the number of features. Let's reshape our data so that it matches the requirements of LSTM.

Execute the following script:

```
train_features =
np.reshape(train_features,
(train_features.shape[0],
train_features.shape[1], 1))
```

In the script above we use the reshape function from the numpy array. The first parameter to the reshape function is the dataset itself. We passed it the training_feature list. The second parameter is the number of records in the training set. To find the number of records in the training_shape we use the shape attribute. The third feature is the number of time steps which is basically the columns in the training_features list and finally, the last parameter is the number of indicators. Since here we are only using 1 feature i.e. open, we set the number of indicator value to 1.

Creating RNN (LSTM)

We have now converted our data into the right format. The next step is to build our RNN. We will create sequential RNN with multiple layers. It will have LSTM layer and finally a densely connected neural network layer at the end. We also use drop out regularization to avoid over-fitting our data. Execute the following script to import the libraries required to create RNN.

```
from keras.models import Sequential
from keras.layers import Dense
from keras.layers import LSTM
from keras.layers import Dropout

# Initialising the RNN
```

```
regressor = Sequential()
```

In the script above we imported the Sequential class from Keras.models library and *Dense, LSTM* and *Dropout* classes from *Keras.layers library.*

Creating a Sequential Model

We begin by creating a sequential model. We will add layers to this model. Execute the following script:

```
stock_predictor = Sequential()
```

Adding LSTM Layer and Drop Out

We will start building our sequential model with LSTM layer. To add the first LSTM layer, execute the following script:

```
stock_predictor .add(LSTM(units =
50, return_sequences = True,
input_shape =
(train_features.shape[1], 1)))
```

To the add LSTM layer to the sequential model, we pass the LSTM object to the add method. The first parameter to the LSTM object is the number of neurons in the layer, denoted by units. The second parameter is the *return_sequence* which specifies that there will be more LSTM layers stacked on this layer. Finally, the third parameter is the input_shape which takes two arguments: number of time steps and the number of indicators.

After adding the LSTM layer, we need to add regularization which basically refers to the number of neurons dropped during each iteration of training. This is done in order to avoid over-fitting. Execute the following script to add drop out regularization:

```
stock_predictor .add(Dropout(0.2))
```

Here 0.2 means that 20% of the neurons will be dropped during each epoch.

Let's add three more layers of LSTM and dropout regularization. Take a look at the following script

```
stock_predictor .add(LSTM(units = 50, return_sequences = True))
stock_predictor .add(Dropout(0.2))

stock_predictor .add(LSTM(units = 50, return_sequences = True))
stock_predictor .add(Dropout(0.2))
```

```
stock_predictor .add(LSTM(units =
50))
stock_predictor .add(Dropout(0.2))
```

You can see from the last LSTM layer that here we are not using "return_sequence = True" since we are not adding any more LSTM layers after that.

Adding Dense Layer

We created LSTM layers in the last section. However, to make our RNN more robust we will add a dense layer at the end as we did with the CNN. Execute the following script to add a dense layer to our RNN:

```
stock_predictor .add(Dense(units =
1))
```

The number of neurons in the output densely connected layer will be 1 here since we only have to predict a single value i.e. the opening stock price on a given day.

Compiling RNN

Like any other neural network, we need to compile our RNN in order to make train it on the training data. The following script compiles the algorithm.

```
stock_predictor.compile(optimizer
= 'adam', loss =
'mean_squared_error')
```

To compile the algorithm, we need to call the compile method on the Sequential model object which is "stock_predictor" in our case. Here to calculate the loss we will use "mean_squared_error' function and to reduce the loss or to optimize the algorithm, we will use the "adam" optimizer.

Training the Algorithm

The final step is to train the RNN. To do so, we will use the fit method of the Sequential object and pass it train_feature, train_labels, number of epochs and the batch size as shown below:

```
stock_predictor
.fit(train_features, train_labels,
epochs = 100, batch_size = 32)
```

When you execute the above script, you will see that our RNN will start training. Depending on the hardware, the training can take quite a bit of time.

Testing and Making Predictions

To test and evaluate the algorithm that we just trained, we need to perform two steps: First, we have to predict the opening stock prices for January 2017 and then we have to compare the predicted stock prices with the actual stock prices.

Let's first load our test set. Remember the test set has been downloaded and stored in "ms-testing.csv" file. Execute the following script to load the test set:

```
stock_data_testing =
pd.read_csv('D:\Datasets\ms-testing.csv')
stock_testing=
stock_data_testing.iloc[:,
1:2].values
```

Let's see how the trend of our test data for the month of January 2017 which we have to predict. Take a look at the following graph:

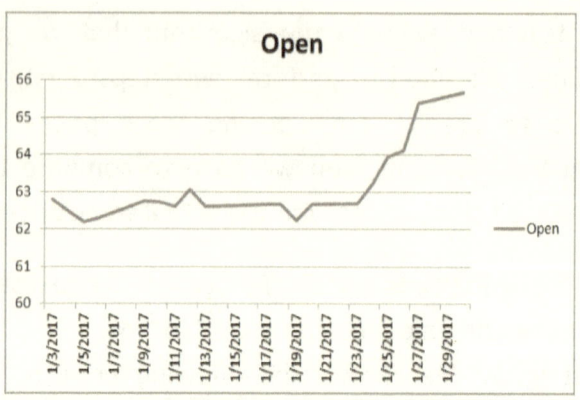

Figure 9.2: Trend of Opening Stock Price for January 2017

From Figure 9.2, we can clearly see that opening stock price of Microsoft for the month of January sees a varying trend at the beginning of January and it ends up on the high. The trend is extremely non-linear and this is why we employed RNN with multiple layers to train the algorithm.

As a next step, we will predict the trends for the month of January and will see if we get a curve similar to the one mentioned in Figure 9.2.

It is very important to mention here that we trained our algorithm by using the opening stock price of previous 60 days to predict the opening stock price of the 61st day.

Now to make predictions on the test set, we again need the data in the same format. For example, to predict the start price for 1st January 2017, we need the opening stock price of last 60 days which will come from training set since training set contains stock prices from January 2012 to December 2016. And similarly, to predict the stock price for 30th January 2017, we will need input from the test set as well as from the training set.

We prepare our input data for predictions by concatenating our training and test sets.

```
stock_total =
pd.concat((stock_data['Open'],
stock_data_testing['Open']), axis
= 0)
```

Since we only need the opening data, therefore we only concatenated the data for the "Open" column. The axis here specifies that the data will be connected vertically along the rows.

Now what we need in the input is the previous 60 stock prices of Microsoft from the first day of January 2017 to the last day of January 2017. If you look at the "ms-training.csv" file, you will see that the first day of January 2017 is the 3rd of January and the last day of January 2017 is the 30th of

January. Therefore we need to iterate through these dates.

However, to get the previous 60 records, we can subtract the length of stock_data_testing from the length of stock_total and subtract 60 from it. This will give us previous 60 stock prices for 3rd January 2017. This will be the lower bound of the input data. For the upper bound we can take the index of the last value of stock_total. Execute the following script:

```
test_inputs = stock_total[len(stock_total) - len(stock_data_testing) - 60:].values
```

Now we need to reshape the numpy array so that it is in the right format and then we will scale it as we did with our training data. To do so, execute the following script:

```
test_inputs = test_inputs.reshape(-1,1)
test_inputs = normalizer.transform(test_inputs)
```

Once the data is scaled, we are ready to create the test data that we will be used for prediction. Our test data will contain 20 records, 1 each for each date in January 2017. Each record will be a feature vector consisting of 60 previous opening stock prices. Execute the following script to create the feature vector:

```
test_features = []
for i in range(60, 80):
test_features.append(test_inputs[i-60:i, 0])
```

Finally, we need to convert our test_features list into numpy array which will then be used to convert into a format that can be used as input to the RNN as we did in the training phase. The following script converts the data into the required format:

```
test_features =
np.array(test_features)
test_features =
np.reshape(test_features,
(test_features.shape[0],
test_features.shape[1], 1))
```

Finally, to predict the stock prices, we will call the predict method on the stock_predictor RNN that we trained in the last section as shown below:

```
predictions
=stock_predictor.predict(test_feat
ures)
```

These predictions are scaled. To get the actual predictions, we need to reverse the scaling, we will use the inverse_transform function of the normalizer object that we created in the last section:

```
predictions =
normalizer.inverse_transform(predi
ctions)
```

Evaluating the Algorithm

The last and final step is to evaluate the algorithm. To do so, we will plot the actual trend of Microsoft prices vs the Predicted trend and see if we get similar curves or not. Execute the following script to do so:

```
plt.figure(figsize=(10,6))
plt.plot(stock_testing, color =
'blue', label = 'Actual Microsoft
Stock Price')
```

```
plt.plot(predictions , color =
'red', label = 'Predicted
Microsoft Stock Price')
plt.title('Microsoft Stock Price
Prediction')
plt.xlabel('Date')
plt.ylabel('Microsoft Stock
Price')
plt.legend()
plt.show()
```

The output looks like this:

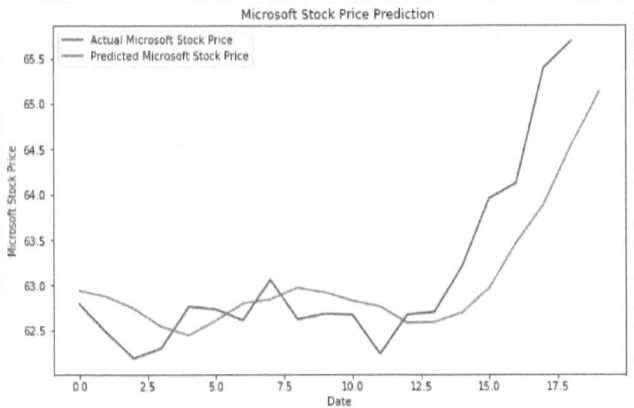

The blue curve is the actual opening day price for our stock for the month of January while the red curve is the predicted opening day price of the stock. You can clearly see that our algorithm has successfully predicted the stock market prices for the month of January. It has not only captured the fluctuating trend at the beginning of January but has also correctly predicted the spike at the end of January.

Conclusion

In this chapter, we implemented an RNN via Python's Keras library. We saw the various steps involved in the implementation of RNN by solving a time series analysis problem where we predicted future stock prices for Microsoft. In the next chapter, we will see the application of RNN in the domain of natural language processing.

Chapter 10

Natural Language Processing with RNN (LSTM)

In the previous chapter, we saw how a Recurrent Neural Network (RNN) can be used to solve time series analysis problems, such as stock price prediction. In this article, we will see the application of RNN, particularly Long Short-Term Memory (LSTM) for natural language processing. Since text consists of sentences, which are basically a sequence of words, LSTM can be used to perform a variety of Natural Language Processing tasks such sentimental analysis, text classification, neural machine translation and so on. In this chapter, we will see how LSTM can be used for text classification. In the next chapter, we will see the application of LSTM for neural machine translation tasks.

Text Classification using RNN (LSTM)

A text classification problem is a type of problems where you are given a text string, which can be a user comment on the Facebook or Twitter or an email message etc. Your job is to classify the text into one of the predefined categories. For instance, for the Facebook comments, predefined categories can be "Positive Comment" and "Negative Comment". Similarly, for the email classification, possible categories can be "Ham" or "Spam". An email is considered "Ham" if it is genuine. On the flip side, if an email is not genuine, for instance, it congratulates you on winning a million dollar lottery; it probably is a spam email.

Text classification is a supervised learning problem, where you are given a set of documents as well as the category to which the document belongs. Your task is to train the algorithm on this data and predict the category for an unseen document. A document can be as long as a full book or an article, or as short as a single sentence.

Several machine learning models have previously been developed for tasks like text classification. However, deep learning techniques have proven to be more successful than traditional machine learning Techniques. Particularly, LSTM has been proven as the state of the art for text classification problems. LSTM is a type of recurrent neural network, which we studied in detail in Chapter 8. We used LSTM for time series analysis in Chapter 9. However, we did not dig deeper into the architecture of the LSTM. In this chapter, we will use LSTM for text classification. But before that, we will study the architecture of LSTM in detail.

LSTM Architecture

Let's first revise the architecture of RNN. The basic principle of RNN is that at time t, it receives the input Xt and the output at the previous time t-1.

Note: Images for this section have been taken from Colah's Blog. It is a very good resource to understand LSTM.

RNN Revision

The basic architecture of RNN looks like this:

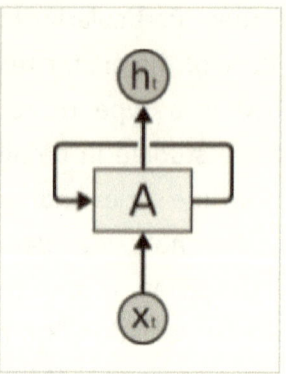

This is an LSTM with one node. It has one input represented by X_t, one activation function that acts on the dot product of weights and one output, known as h_t. If you unroll this node in a temporal space, it looks likes the following figure.

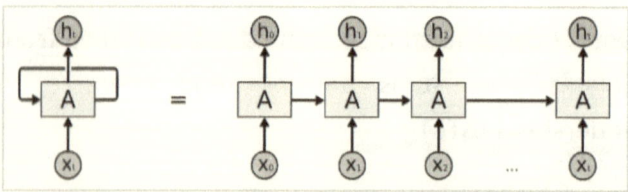

Here x is the input and h is the output. Let's start from time t=0. At t=0, the input X_0 is passed through the node and the output h_0 is received. At time step t=1, the input x_1 along with the previous output h_0 is passed as input to the next node. The process continues until the final output is received. This architecture of neural network makes them ideal for the sequence problems such as time series analysis that we saw in the last chapter.

The feedback mechanism enables RNN to take into account, past information while making decisions. RNN architecture has plenty of applications such as text classification, image captioning, speech recognition and neural machine translation. However, there is a problem with this simple RNN architecture.

Problem with Simple RNN

Though theoretically, RNN is capable of retaining the past information, in practice, this is not the case. RNN is only capable of remembering the recent information. It is not capable of remembering information after a large gap of sequence steps. In some cases, this is not a problem. For instance, take a look at the sentence "Birds fly in the ___". Our RNN can easily predict the next word which can be "Sky" since the past 4 words were enough to predict the next word.

However, in cases where the information from previous steps that are far away from the current time step, is needed, the RNN doesn't seem to perform correctly. For instance, consider the following paragraph. The dotted lines represent any random text.

"My father is an Italian. He likes to eat Pasta and Pizza. He watches football .... and he speaks ____ ".

Now if an RNN wants to predict the next word after "speaks", it needs information from the beginning of the sentence i.e "My father is an Italian". Unfortunately as the gap between the current time step and the previous time steps increase, RNN starts to forget the previous information due to phenomena called vanishing gradients. This is called a problem of long-term dependency.

Solution: LSTM

The LSTM (Long Short-Term Memory) neural network is capable of retaining information learned through a long number of time steps and hence can solve the problem of long-term dependency. Developed by [Hochreiter&Schmidhuber (1997)](), several variants of LSTM have been developed till now. However, we will study the basic version in this section.

The memory module of RNN consists of one layer. Where one input, along with the previous output or hidden state is being fed into the next node. The architecture of LSTM differs from RNN in a way that in addition to the output, it has a cell state vector as well which is responsible for maintaining long-term memory. The basic architecture of LSTM is shown below:

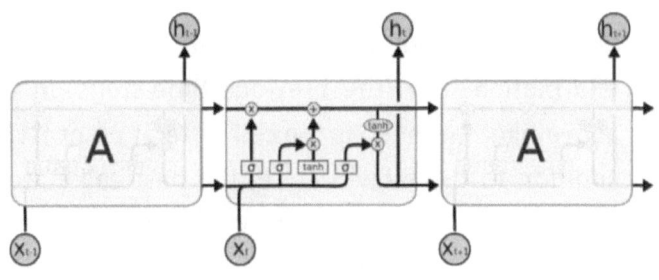

Don't be intimidated by the above architecture, we will see each component of LSTM in detail.

The first component of an LSTM is the cell state. Take a look at the following figure:

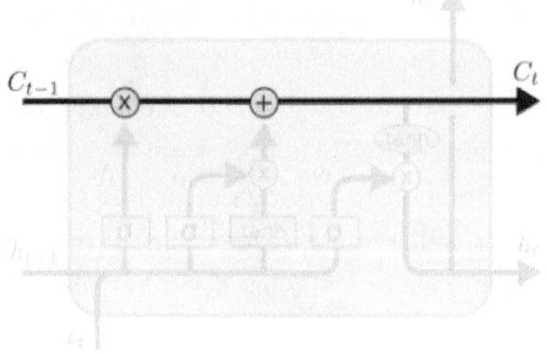

Cell state contains values that store the information from the previous time-steps. In the above figure, the C_{t-1} is the cell state at point t-1. In addition to the hidden state, the past cell state is also passed as input to the LSTM at time step t. Similarly, the cell state at time t i.e. C_t is passed as input to the LSTM at time t+1 and so on.

However, we do not need all of the past information. And depending upon the input we need to update the past information. To remove, add or modify information in the cell state, different types of gates are used in LSTM. Gates consist of a sigmoid operation in the neural net layer, followed by a pointwise multiplication operation.

The following operations are performed in an LSTM neural network node at each time step.

1. In the first step, the information from the previous time step that is needed to be forgotten is passed to the cell state. The gate that performs this operation is called forget gate. Take a look at the following figure. Here f_t refers to the forget gate.

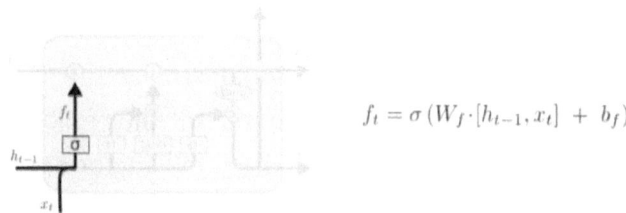

$$f_t = \sigma\left(W_f \cdot [h_{t-1}, x_t] + b_f\right)$$

In the forget gate, the previous hidden state, and the current input are concatenated and passed through a sigmoid function. For each value in the cell state, the sigmoid function returns 1 or 0. Here 0 represents that okay forget this, while 1 represents that keep this information. The result of the sigmoid operation is pointwise multiplied with the existing cell state.

Real world example of the forget gate can be the gender of the person. If a new person with a different gender comes, we need to change all the pronouns. The forget gate will help us forget the gender of the previous person.

2. The next step is to find out the new information (if any, that is to be stored in the cell state). To find out the new information to be updated, two operations are required to be performed. The first one is to identify what part of the cell state is to be updated. This is defined by the input gate represented by i_t in the following figure. The second part is the vector of new values that will replace the old value and is represented by $\tilde{C}_t$ as shown in the following figure.

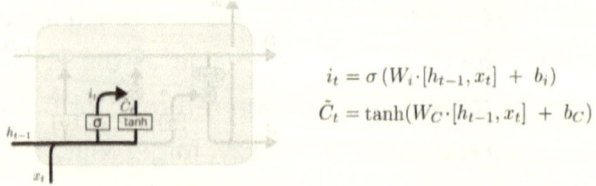

$$i_t = \sigma(W_i \cdot [h_{t-1}, x_t] + b_i)$$
$$\tilde{C}_t = \tanh(W_C \cdot [h_{t-1}, x_t] + b_C)$$

A real-world example of this step is that i_t will find that gender has to be updated, while the $\tilde{C}_t$ will result in the new value of the gender.

3. Now we have all the information needed to update the cell state. We just need to perform the operations that we derived in step 1 and step 2. Step 3 can be visually represented as:

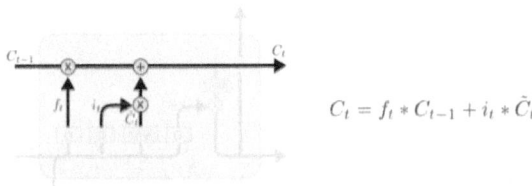

$$C_t = f_t * C_{t-1} + i_t * \tilde{C}_t$$

4. We calculated the cell state; the next step is to find the output or the hidden state for the next state. It again has two parts; the first part is the sigmoid of the input and the previous hidden layer. This decides the part of the state that we are going to output. The cell state is passed through "tan" activation function so that the output vector is between 0 and 1 and then multiplied by the result of the sigmoid in order to output only the part defined by the sigmoid operation. The operations performed for the calculation of the output have been shown below:

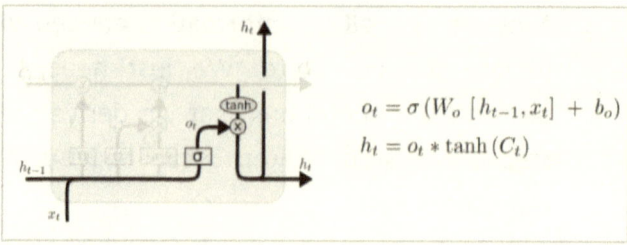

$$o_t = \sigma\left(W_o\left[h_{t-1}, x_t\right] + b_o\right)$$
$$h_t = o_t * \tanh\left(C_t\right)$$

LSTM in Practice

Let's take a look at a very simple of LSTM to understand the concept of cell state, hidden state, and the output. Take a look at the following script:

```
from keras.models import Model
from keras.layers import Input, LSTM
import numpy as np
importmatplotlib.pyplot as plt


sequence_length = 8
features = 2
lstm_nodes = 3


X = np.random.randn(1, sequence_length, features)
```

```
input_ =
Input(shape=(sequence_length,
features))
rnn = LSTM(lstm_nodes)
output = rnn(input_)

model = Model(inputs=input_,
outputs = output)
out = model.predict(X)
print("out:", out)
```

In the script above, we create a simple LSTM with three nodes. The input to the LSTM is a single sequence of length 8 and 2 features per element in the sequence. We use the Keras Functional API to create our LSTM model.

In our LSTM, we first need to create the Input layer and specify the shape of the input. The Input layer takes two parameters: The length of the sequence and the number of features per element in the sequence.

Next, we create a simple LSTM and pass it the number of nodes. Finally, for output, we can simply use the input passed through output.

Next, we create our model. The model class again takes two parameters, the input, and the output. We can now use this model to directly predict the output of the sequence using the "predict" function. The data point that we generated is passed as input to "predict". In the output, you will see a list of three numbers. Each number corresponds to the output produced by one weight. The output looks like this:

```
out: [[-0.09068623  0.17241372  0.04843738]]
```

Using Return_State Parameter

You can also see the value of the cell state and the hidden state by passing *"return_state = True"* to the LSTM function. Modify the above script as follows:

```
from keras.models import Model
from keras.layers import Input, LSTM
import numpy as np
importmatplotlib.pyplot as plt

sequence_length = 8
features = 2
lstm_nodes = 3
```

```python
X = np.random.randn(1, sequence_length, features)


input_ = Input(shape=(sequence_length, features))
rnn = LSTM(lstm_nodes, return_state= True)
output = rnn(input_)


model = Model(inputs=input_, outputs = output)
out, hidden, cell = model.predict(X)
print("out:", out)
print("hidden:", hidden)
print("cell:", cell)
```

The output looks like this:

```
out: [[ 0.10416643 -0.13675779  0.1200356 ]]
hidden: [[ 0.10416643 -0.13675779  0.1200356 ]]
cell: [[ 0.2025083  -0.4116366   0.33810562]]
```

You can see that with *"return_state = True"*, we can get the value of the hidden state, output and cell state at the end of the sequence. Another interesting, observation here is that the output is actually same as the value of the hidden state at the end of the sequence. This shows that depending upon the node or the time step, the output is actually the hidden state.

Using Return_Sequences Parameter

In addition to finding the output at the end of the sequence, you can also calculate the output at each time step of the sequence. To do so, you need to pass the "return_sequences = True" to the LSTM class as shown below. Take a look at the following script:

```
from keras.models import Model
from keras.layers import Input, LSTM
import numpy as np
importmatplotlib.pyplot as plt

sequence_length = 8
features = 2
lstm_nodes = 3


X = np.random.randn(1, sequence_length, features)
```

```python
input_ = 
Input(shape=(sequence_length, 
features))

rnn = LSTM(lstm_nodes, 
return_state= True, 
return_sequences = True)

output = rnn(input_)

model = Model(inputs=input_, 
outputs = output)

out, hidden, cell = 
model.predict(X)

print("out:", out)

print("hidden:", hidden)

print("cell:", cell)
```

The output of the script above looks like this:

```
out: [[[-0.13825381 -0.08070102  0.07761218]
  [-0.28424516 -0.08341846  0.1885149 ]
  [-0.29308954 -0.04668333  0.2016785 ]
  [-0.2653988  -0.12767167  0.07278793]
  [-0.19219656 -0.12255747 -0.09975541]
  [-0.20349663 -0.21057348 -0.18972561]
  [-0.08813298 -0.12625995 -0.2284562 ]
  [ 0.0885969  -0.01841614 -0.25060138]]]
 hidden: [[ 0.0885969  -0.01841614 -0.25060138]]
 cell: [[ 0.15074036 -0.05491044 -0.65512824]]
```

In the script above you can see 8 output values for 3 weights. We had 8 elements in the sequence and since we set "return_sequences = True", we got 8x3 output values, one each for time step and corresponding weight.

You can also see the hidden state and the cell state at the end of the sequence. Here you can again observe that the output at the final time step is equal to the hidden state values.

Spam and Ham Email Classification

In the previous section, we saw the theory behind LSTM in detail. In this section, we will see the application of LSTM for ham and spam email classification. Since LSTM is ideal for sequence problems, therefore we can use it for natural language processing applications such as text classification.

The dataset for this application can be found in the Datasets folder. The path to the datasets folder has been mentioned in the introduction section of the book. The dataset that we are going to use for this example is the "spam.csv"

We will follow the standard machine learning library for this task. Follow these steps:

1. Importing Required Libraries

The first step is to import the required libraries for this task. Execute the following script to do so:

```
import os
import sys
import numpy as np
import pandas as pd
importmatplotlib.pyplot as plt

fromkeras.models import Model
fromkeras.layers import Dense, Embedding, Input
fromkeras.layers import LSTM, Bidirectional, GlobalMaxPool1D, Dropout
fromkeras.preprocessing.text import Tokenizer
fromkeras.preprocessing.sequence import pad_sequences
fromkeras.optimizers import Adam
fromsklearn.metrics import roc_auc_score
fromsklearn.preprocessing import LabelEncoder
```

2. Configuration Settings

Let's set up the configurations for our application. Execute the following script:

```
MAX_SENTENCE_LENGTH = 150
VOCABULARY_SIZE = 2000
VECTOR_DIM = 100
TRAIN_TEST_SPLIT = 0.2
BATCH_SIZE = 64
EPOCHS = 10
```

Here MAX_SENTENCE_LENGTH will be the maximum length of the sequence. Since our sentences are of variable size, but we can only input fixed size data to LSTM, we will pad the shorter sentences by zero and make them the same length as the largest sentence of the corpus.

VOCABULAR_SIZE is the number of words that we will use to create the embedding layer. The VECTOR_DIM will be the number of features per word in the sentence. The TRAIN_TEST_SPLIT will define the size of the training and test sets. A value of 0.2 here indicates that our data will be divided into 20% test set and 80% train set. Our batch size will be 64 and we will have 10 epochs.

3. Loading Pre-trained Word Embeddings

Natural language consists of texts. However, machine learning algorithms work with numbers. Therefore, we need to convert text into numbers. Text representation in numerical format is called embeddings.

There are several ways to convert text into numbers. We can create embeddings from scratch or we can use pre-trained word embeddings. For the sake of this chapter, we will use pre-trained word embeddings called Glove, developed by researchers at Stanford University. The text file for word embedding can be found in the datasets folder by the name "glove.6b.100d". In this file, 400000 words have been represented as 100-dimensional vectors.

Next, we need to load these word vectors into our application. Execute the following script to do so. Please double check the path of the "glove.6b.100d" file before executing the following script:

```
print('Loading Pretrained Word Vectors')
embedding_dic = {}
```

```
with
open("E:/Datasets/glove.6b.100d.tx
t", encoding= "utf8") as
vector_list:

for line in vector_list:
dimensions = line.split()
word = dimensions[0]
vector =
np.asarray(dimensions[1:],
dtype='float32')
embedding_dic[word] = vector

print("Total Embedding Vectors:" +
str(len(embedding_dic)))
```

4. Importing the Dataset

Let's import the dataset for this chapter. Execute the following script to import the data and display the first five records:

```
print('Read mails')
email_data =
pd.read_csv("E:/Datasets/spam.csv"
, delimiter=',',encoding='latin-
1')
email_data.head()
```

The first five records look like this:

	v1	v2	Unnamed: 2	Unnamed: 3	Unnamed: 4
0	ham	Go until jurong point, crazy.. Available only ...	NaN	NaN	NaN
1	ham	Ok lar... Joking wif u oni...	NaN	NaN	NaN
2	spam	Free entry in 2 a wkly comp to win FA Cup fina...	NaN	NaN	NaN
3	ham	U dun say so early hor... U c already then say...	NaN	NaN	NaN
4	ham	Nah I don't think he goes to usf, he lives aro...	NaN	NaN	NaN

The first column i.e. v1 contains the labels or the classes for the text while the second column contains messages. We do not need the last three columns. Let's drop these columns. Execute the following script:

```
email_data.drop(['Unnamed: 2',
'Unnamed: 3', 'Unnamed:
4'],axis=1,inplace=True)
email_data.info()
```

5. Text Preprocessing

Let's create two separate variables for emails and their corresponding labels. Execute the following script to do so:

```
emails =
email_data["v2"].fillna("DUMMY_VAL
UE").values
labels = email_data["v1"]
```

Let's first convert our labels to numbers. We will use LabelEncoder class for that, which converts text labels into corresponding integers. Execute the following script:

```
le = LabelEncoder()
labels = le.fit_transform(labels)
labels =labels.reshape(-1,1)
```

After executing the above script if you compare the v1 column of the email_datadataframe and the labels list, you can see clear corresponding between the label name and its corresponding integer encoding as shown in the following figure:

email_data - DataFrame		labels - NumPy array
Index	v1	0
0	ham	0
1	ham	0
2	spam	1
3	ham	0
4	ham	0
5	spam	1
6	ham	0
7	ham	0
8	spam	1
9	spam	1
10	ham	0
11	spam	1
12	spam	1
13	ham	0

Like labels, we need to convert our email messages into numeric format where each sentence is of same length as the maximum sequence length. The following script performs that task.

```
tok = Tokenizer(num_words=VOCABULARY_SIZE)
tok.fit_on_texts(emails)
email_sequences = tok.texts_to_sequences(emails)
email_sequence_matrix = pad_sequences(email_sequences, maxlen = MAX_SENTENCE_LENGTH)
```

Here we use the Tokenizer class, to tokenize sentences into words. We then used the "texts_to_sequence" method of the Tokenizer object to convert words into integers. At this point time, sentences have been converted into corresponding integers but they are still not of the same size. To make sentences have the same size as the size of the largest sentence, we use the pad_sequence class.

6. Create Embedding Matrix

We loaded our pre-trained embeddings and we converted our emails into integers. Now is the time to create embeddings for the words in our email messages.

If you execute the following script, you will see that our dataset has 8920 unique tokens or words:

```
word2idx = tok.word_index
print('Found %s unique tokens.' % len(word2idx))
```

Here "word2idx" basically contains mappings between words and their corresponding integer values.

Let's find the number of words that we want in our vocabulary. Earlier, we set this number to 2000. However, if the unique tokens are less than 2000, we will set our vocabulary size to unique tokens. The following script does that:

```
total_words = min(VOCABULARY_SIZE, len(word2idx) + 1)
```

We will then create an empty embedding matrix of size total_words x VECTOR_DIM as shown below:

```
embedding_matrix = np.zeros((total_words, VECTOR_DIM))
```

Finally, we will fill this embedding matrix, with the vectors of 2000 most occurring words as shown below:

```
for word, index in
word2idx.items():
if index < VOCABULARY_SIZE:
vector = embedding_dic.get(word)
if vector is not None:
embedding_matrix[index] = vector
```

7. Create Embedding Layer

After creating the embedding matrix, the next step is to create the embedding layer. Embedding layer basically tells our LSTM the total vocabulary size, the dimensions of each element in the sequence, pre-trained embedding matrix to use (if any) and the maximum length of the sequence. The following script creates the embedding layer for our script:

```
embedding_layer = Embedding(
total_words,
   VECTOR_DIM,
weights=[embedding_matrix],
input_length=MAX_SENTENCE_LENGTH,
```

```
trainable=False
)
```

8. Training The Model

We have performed all the preprocessing required to train the model. Now is the time to actually train our model. Execute the following script to do so:

```
input_ = Input(shape=(MAX_SENTENCE_LENGTH,))
x = embedding_layer(input_)
x = LSTM(20)(x)

output = Dense(1, activation="sigmoid")(x)

model = Model(input_, output)
model.compile(
loss='binary_crossentropy',
optimizer=Adam(lr=0.01),
metrics=['accuracy']
)
```

We have one LSTM in our model with 20 nodes. We also have one dense layer. The activation function that we are going to use is "sigmoid" function while the loss function is "binary_crossentropy" since this is a binary classification problem.

Now as the last step, let's train our model on our data and see the accuracy returned by our algorithm. Execute the following script:

```
print('Training model...')
classifier = model.fit(
email_sequence_matrix,
labels,
batch_size=BATCH_SIZE,
epochs=EPOCHS,
validation_split =
TRAIN_TEST_SPLIT
)
```

At the end of the execution, the algorithm reached an accuracy of 99.44% which is brilliant. The final result is shown below:

```
Epoch 7/10
4457/4457 [==============================] - 4s 959us/step - loss: 0.0261 - acc:
0.9928 - val_loss: 0.0534 - val_acc: 0.9874
Epoch 8/10
4457/4457 [==============================] - 4s 959us/step - loss: 0.0176 - acc:
0.9955 - val_loss: 0.0465 - val_acc: 0.9857
Epoch 9/10
4457/4457 [==============================] - 4s 960us/step - loss: 0.0146 - acc:
0.9962 - val_loss: 0.0364 - val_acc: 0.9901
Epoch 10/10
4457/4457 [==============================] - 5s 1ms/step - loss: 0.0191 - acc:
0.9944 - val_loss: 0.0547 - val_acc: 0.9821
```

Finally, let's plot the accuracy of the algorithm against each epoch. Execute the following script to do so:

```
plt.plot(classifier.history['loss'], label='loss')
plt.plot(classifier.history['val_loss'], label='val_loss')
plt.legend()
plt.show()
```

The output looks like this:

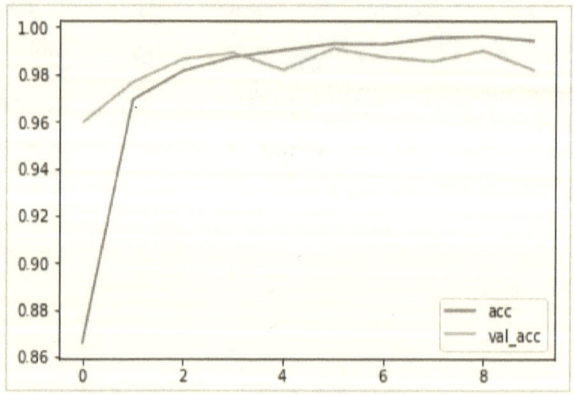

You can see that the accuracy converges after the 6th epoch. Also, both accuracy and value accuracy is almost the same which shows that our algorithm did not over-fit.

Conclusion

In this chapter, we study LSTM in detail. We saw how LSTM can be used in natural language processing. We used LSTM to solve a simple text classification task where we classified email messages into ham and spam categories. In the next Chapter, we will see the application of LSTM for neural machine translation tasks. We will develop a very simple Chabot capable of replying to basic queries.

Chapter 11

The sequence to Sequence Modelling with LSTM

In the previous chapter, we saw how LSTM can be used for text classification. However, LSTM can be used for a variety of natural language processing tasks. Sequence to sequence, also known as seq2seq is one of the most famous examples of the LSTM applications for NLP. In this article, we will first study the theory behind seq2seq models. We will then move on to build a very simple Chatbot, capable of answering basic user queries. The Chatbot will not be very advanced, but you will get the idea behind the working principles of seq2seq architecture.

What is Sequence to Sequence Modelling?

In chapter 8, we studied that RNN can be used for a variety of applications. The applications of RNN can be divided into the following types:

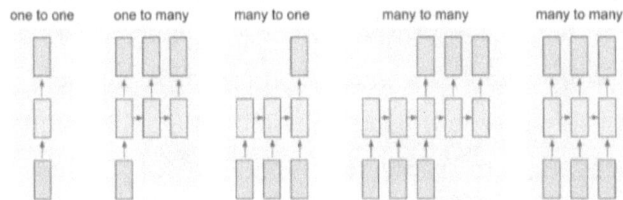

An example of "one to one" application can be an Image labeling task where you have an image as input and its corresponding label as output. Similarly, the example of RNN "one to many" applications can be image captioning where a single image is an input and a sentence explaining the image is the output. Text classification is a classic example of "many to one" applications.

Finally, we have two types of "many to many" applications. In the first type, inputs and outputs have varying size while in the second type the input and outs have the same size. These "many to many" applications are solved using the sequence to sequence technique that we are going to learn in this chapter.

Parts of speech tagging is an example of "many to many" applications with a fixed size. For each word in the output, you have a corresponding word in the output. On the other hand, Neural Machine Translation is an example of "many to many" applications with varying input and output sizes. For instance, a sentence in one language may have 4 words, while its translation can have 3 words. In this Chapter, our focus will be on the application of sequence to sequence modeling for "many to many" applications with varying input and output size.

Sequence to Sequence Architecture

Seq2seq model executes in two phases: Training and Inference. The architecture consists of two stacked LSTM networks called encoder and decoder. Take a look at the following figure for a high-level understanding of Sequence to Sequence Architecture:

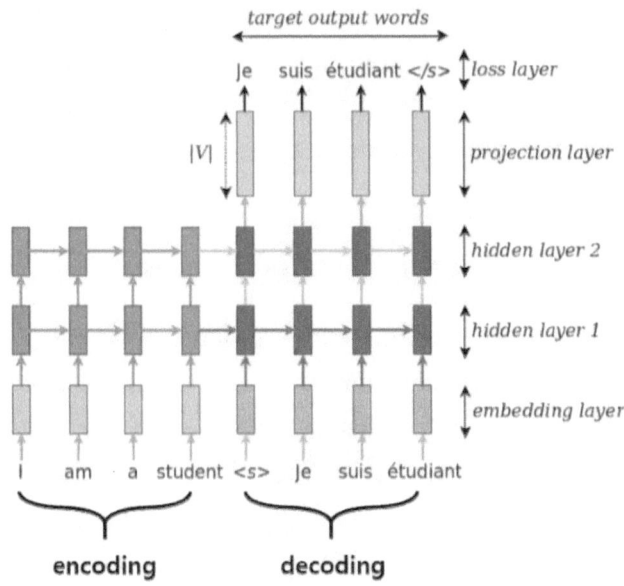

Encoder Decoder Model for Training Seq2Seq
(Image Reference)

As you can see in the figure, the encoder is basically an LSTM, which is responsible for encoding input sequence and producing thought vector. Though vector consists of the hidden state and the cell state that will be used as input by the decoding LSTM. The encoder LSTM node layers have been represented by the dark blue rectangles in the above figure. The output from the Encoding LSTM is used as the input to the decoding LSTM (Shown in dark red). However decoding LSTM not only receives the input from encoding LSTM, but it also receives the output, moved one step forward as the input to generate the final response. Let's take a look at the steps being performed during the executing of Sequence to sequence modeling, in detail.

Step 1 (Preparing Inputs for Encoding LSTM)

The input to the Encoding LSTM will be an input sentence, converted into the embedding vector. Remember, in the last chapter, we used Glove embedding vectors, in order to convert words into 100-dimensional embedding vectors. The output from the encoding LSTM will be the hidden state and cell state. We discussed the concepts of hidden state and cell state in the last chapter.

Step 2 (Preparing Inputs for Decoding LSTM)

The decoder LSTM receives two inputs from the encoding LSTM along with target text offset by one element as shown in the following figure:

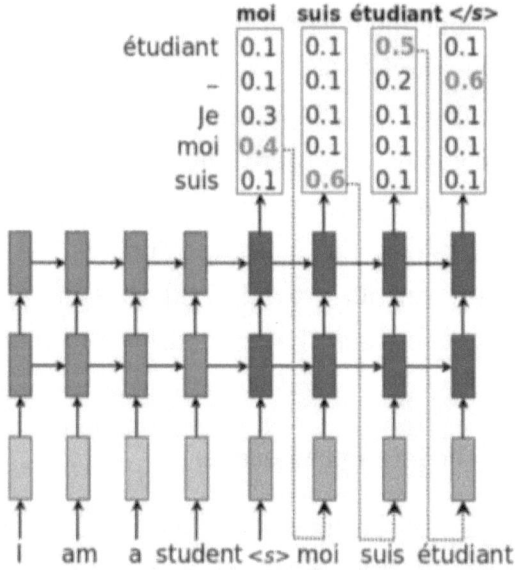

Decoder Inputs (Image Reference)

You can see from the figure above that the decoder model receives the hidden state and cell state of the encoding LSTM as well as the target output as input, offset by one. You can see in the above figure that our target is represented as one hot encoded string of all the words in the vocabulary, while the input is the same string in its original format.

Step 3 (Training the Encoder Decoder Model)

Once the inputs and outputs have been decided for the encoder and decoder model, the next step is to train the model. Since our output is actually a matrix where rows represent the sequence length while columns represent the one hot encoded output for each element in the sequence, we will use softmax as the activation function. In every iteration, weights are adjusted to minimize the difference between the input string and predicted string.

Step 4 (The Inference Decoder)

Once the model is trained, the next step is to make predictions. It is important to note that during the training phase we had all the elements in the input sequence, however for the outputs, this is not going to be the case since we do not know every element in the input beforehand. But if you closely look at the training phase, you can see that the input to the next unit in the decoder LSTM is basically the output from the previous unit. This is shown in the following figure:

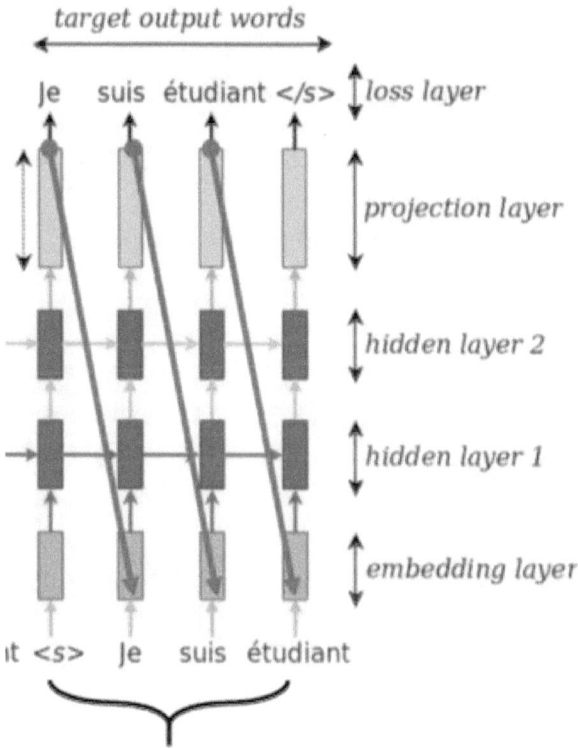

Decoder Inputs during Inference/Prediction Phase
(Image Reference)

Step 5 (Convert Embeddings to Words)

The predictions are in the form of embedding vectors which we have to convert back to the words. This is the final phase of the sequence to sequence model.

The steps explained in this section will become clear in the next section when we will implement the sequence to sequence model in code.

Creating a Chatbot Using Sequence to Sequence Model

We have covered the theory behind sequence to sequence architecture. Now is the time to see a real-world application of sequence to sequence modeling.

In this section, we will use Python's Keras library to develop a sequence to sequence chatbot. The code template for the sequence to sequence model is available at [Keras official blog](). This code can be used to create any sequence to sequence model. We will use this code to create our Chabot. So, let's start.

Step1: Importing the Library

As always, the first step is to import libraries required to implement the sequence to sequence model. Execute the following script:

```
import os, sys
import re
```

```python
from keras.models import Model
from keras.layers import Input, LSTM, GRU, Dense, Embedding
from keras.preprocessing.text import Tokenizer
from keras.preprocessing.sequence import pad_sequences
from keras.utils import to_categorical
import numpy as np
import matplotlib.pyplot as plt
```

Step 2: Configuration Settings

```
BATCH_SIZE = 4
EPOCHS = 20
LSTM_NODES = 128
MAX_SENTENCE_LENGTH = 200
VOCABULARY_SIZE = 2000
VECTOR_DIM = 100
```

Here MAX_SENTENCE_LENGTH will be the maximum length of the sequence. VOCABULAR_SIZE is the number of words that we will use to create the embedding layer. The VECTOR_DIM will be the number of features per word in the sentence. Our batch size will be 4 and we will have 20 epochs.

Step 3: Importing and Preprocessing the Dataset

Now is the time to load and preprocess the data. The dataset that we are going to use is chat conversation between two humans. Where one of the humans play the role of a robot. The files for this section are available in the Datasets folder by the name "human_text.txt" and "robot_text.txt". The dataset can be downloaded online from this kaggle link as well.

In the case of sequence to sequence learning, we need three separate lists. The first one is the input text list that contains the input to the encoder LSTM, the second one is the target text list which will be the output of the decoder LSTM and the third one is the target text input list which will act as input to the decoder LSTM. Let's create these three lists:

```
input_texts = []
target_texts = []
target_texts_inputs = []
```

Next, we will read both the "human_text.txt" and "robot_text.txt", line by line. Remember that each line in the "robot_text.txt" is a response to the corresponding line in the "human_text.txt".

We will read the files line by line, remove all the punctuation and extra spaces from the lines. We will then check if the number of words in the input line is greater than the MAX_SENTENCE_LENGTH, simply ignore the sentence since we do not want very big sentences in our document.

Finally, we will insert lines from "human_text.txt" in the input_texts lists, and the lines from "robot_text.txt" will be inserted in the target_texts lists. We prefix all the lines in target text with the "eos" (end of sentence tag). The lines from the "robot_text.txt", will also be inserted to the target_texts_inputs list, however, they will be prefixed with "sos" tag since the input to the decoder LSTM will be same as the target output, however, it will be offset by one using the "sos" tag.

Take a look at the following script to perform the data import and preprocessing tasks. You will need to change the path as per your directory structure.

```
string = "how old are you    how do you look like    where do you live"
```

```python
formatted = re.sub(r'\s+', ' ', string, flags=re.I)


with open("E:/human_text.txt", encoding="utf8") as textfile1, open("E:/robot_text.txt", encoding="utf8") as textfile2:
    for human_text, robot_text in zip(textfile1, textfile2):

        human_text = human_text.strip()
        robot_text = robot_text.strip()

        human_text = re.sub(r'\W', ' ', str(human_text))
        robot_text = re.sub(r'\W', ' ', str(robot_text))

        human_text = re.sub(r'\s+', ' ',  human_text, flags=re.I)
        robot_text = re.sub(r'\s+', ' ',  robot_text, flags=re.I)

        input_text = human_text
```

```
        target_text = robot_text +
' <eos>'
        target_text_input = '<sos>
' + robot_text



        if len(input_text.split())
> MAX_SENTENCE_LENGTH:
            break;


input_texts.append(input_text)

target_texts.append(target_text)

target_texts_inputs.append(target_text_input)

print("num samples:",
len(input_texts))
```

Let's print the number of samples in our dataset. Execute the following script to do so:

```
print("num samples:",
len(input_texts))
```

The output will display 581. Our sample size is really small but you will see that our chatbot will be able to answer really small queries.

Step 4: Tokenizing and Padding the Sentences

Before we can feed our sentences into LSTM, we need to convert them into word vectors. For Input, we will use glove word embeddings for our data but before that, we need to convert sentences into integers so that the integers can be used to index the embedding matrix, which contains the vector representation of the word. For the target text, we will use Keras built-in embedding, which we will see in a later section. For now, let's tokenize our input and target data. Execute the following script to tokenize the input:

```
tokenizer_inputs =
Tokenizer(num_words=VOCABULARY_SIZ
E )
tokenizer_inputs.fit_on_texts(inpu
t_texts)
input_sequences =
tokenizer_inputs.texts_to_sequence
s(input_texts)

word2idx_inputs =
tokenizer_inputs.word_index
```

```
max_len_input = max(len(s) for s
in input_sequences)
```

In the script above, we tokenize the inputs, find the number of unique tokens in the input data and the length of the longest sentence.

Similarly, to tokenize the target text and the target text input. Execute the following script:

```
tokenizer_outputs =
Tokenizer(num_words=VOCABULARY_SIZ
E , filters='')

tokenizer_outputs.fit_on_texts(tar
get_texts + target_texts_inputs) #
inefficient, oh well

target_sequences =
tokenizer_outputs.texts_to_sequenc
es(target_texts)

target_sequences_inputs =
tokenizer_outputs.texts_to_sequenc
es(target_texts_inputs)



word2idx_outputs =
tokenizer_outputs.word_index

num_words_output =
len(word2idx_outputs) + 1

max_len_target = max(len(s) for s
in target_sequences)
```

Both our inputs and output sentences are not of the same size. We will pad the inputs according to the longest sentence in the input text. Similarly, the target text and target text input will be padded as per the longest sentence in the output data. Execute the following script to pad input and output data:

```
encoder_inputs =
pad_sequences(input_sequences,
maxlen=max_len_input)


decoder_inputs =
pad_sequences(target_sequences_inp
uts, maxlen=max_len_target,
padding='post')

decoder_targets =
pad_sequences(target_sequences,
maxlen=max_len_target,
padding='post')
```

Step 5: Loading Word Embeddings and Creating Word Embedding Matrix

We will use Glove embeddings to achieve vector representation of the word. Execute the following script to load pertained word vectors and create word embeddings:

```python
print('Loading Word Vectors')
embedding_dic = {}
with open("E:/glove.6b.100d.txt", encoding= "utf8") as vector_list:

  for line in vector_list:
    dimensions = line.split()
    word = dimensions[0]
    vector = np.asarray(dimensions[1:], dtype='float32')
    embedding_dic[word] = vector


num_words = min(VOCABULARY_SIZE, len(word2idx_inputs) + 1)
embedding_matrix = np.zeros((num_words, VECTOR_DIM ))
for word, i in word2idx_inputs.items():
  if i < num_words:
    embedding_vector = embedding_dic.get(word)
```

```
    if embedding_vector is not
None:
     embedding_matrix[i] =
embedding_vector
```

Step 6: Creating Input Embedding Layer and Output One Hot Encoded Layer

Before we create our model, we need to create the input embedding layer and the one hot encoded output layer. Since we are using Softmax function at the output node, our output should be one hot encoded. The following script creates an embedding layer and one hot encoded output vector:

```
embedding_layer = Embedding(
  num_words,
  VECTOR_DIM,
  weights=[embedding_matrix],
  input_length=max_len_input,
  # trainable=True
)
```

```python
decoder_targets_one_hot = 
np.zeros(
  (
    len(input_texts),
    max_len_target,
    num_words_output
  ),
  dtype='float32'
)


for i, d in 
enumerate(decoder_targets):
  for t, word in enumerate(d):
    decoder_targets_one_hot[i, t, word] = 1
```

Step 7: Creating the Model

We have preprocessed all the data needed to create our model. Now is the time to create our model.

As we said earlier, we need to create two LSTMs: an encoder LSTM and a decoder LSTM. Let's first define the input, embedding layer and the output for the encoder. Execute the following script:

```
encoder_inputs_placeholder =
Input(shape=(max_len_input,))
x =
embedding_layer(encoder_inputs_pla
ceholder)
encoder = LSTM(
  LSTM_NODES,
  return_state=True,
)
encoder_outputs, h, c = encoder(x)
encoder_states = [h, c]
```

In the script above, we create input place holders for the encoder input. We know that encoder receives a sequence of input sentences. Therefore, the shape of the input is the "max_len_input" parameter.

Next, we pass our input through the embedding layer. Finally, we create our encoder and set its return state to true. The input passed through the embedding layer is passed to the encoder which returns output, hidden state, and cell state. From the encoder, we only need the cell state and hidden state, which we capture in the "encoder_states" list.

Similarly, let's create input and output for the decoder. Execute the following script:

```
decoder_inputs_placeholder =
Input(shape=(max_len_target,))

decoder_embedding =
Embedding(num_words_output,
LSTM_NODES)
decoder_inputs_x =
decoder_embedding(decoder_inputs_p
laceholder)

decoder_lstm = LSTM(
  LSTM_NODES,
  return_sequences=True,
  return_state=True,

)
decoder_outputs, _, _ =
decoder_lstm(
  decoder_inputs_x,
  initial_state=encoder_states
)
```

In the script above, we first define the input place holder for the decoder. Notice the length of the input shape for the input place holder is "max_len_target" which is the length of the longest sentence in target_text list. Next, we pass the input placeholder through embedding layer. We then create an LSTM with return_sequences = True since we need output at every point in the decoder. Notice that the initial_state of the decoder_lstm is set to encoder_states which consists of the hidden state and cell state of the encoder at the output unit.

Next, we need to create a dense layer and pass the decoder output to the dense layer. Take a look at the following script:

```
decoder_dense =
Dense(num_words_output,
activation='softmax')
decoder_outputs =
decoder_dense(decoder_outputs)
```

Next, we need to create the model and compile it as shown below:

```python
model =
Model([encoder_inputs_placeholder,
decoder_inputs_placeholder],
decoder_outputs)

model.compile(
  optimizer='rmsprop',
  loss='categorical_crossentropy',
  metrics=['accuracy']
  )
```

Finally, we need to fit our model on our data that we created in Step 1 to Step 6. Execute the following script to do so:

```python
r = model.fit(
  [encoder_inputs,
decoder_inputs],
decoder_targets_one_hot,
  batch_size=BATCH_SIZE,
  epochs=EPOCHS,
  validation_split=0.2,
)
```

You will see that our model will start training. At the end of 20 epochs, you should see an accuracy of around 87.21% as shown in the following figure:

```
Epoch 17/20
464/464 [==============================] - 19s 42ms/step - loss: 0.6820 - acc: 0.8682 - val_loss: 0.8790 - val_acc: 0.8721
Epoch 18/20
464/464 [==============================] - 20s 43ms/step - loss: 0.6605 - acc: 0.8700 - val_loss: 0.8841 - val_acc: 0.8716
Epoch 19/20
464/464 [==============================] - 20s 44ms/step - loss: 0.6416 - acc: 0.8732 - val_loss: 0.8966 - val_acc: 0.8721
Epoch 20/20
464/464 [==============================] - 20s 44ms/step - loss: 0.6206 - acc: 0.8771 - val_loss: 0.8974 - val_acc: 0.8721
```

Step 8: Making Predictions

We have trained our model; the next step is to make predictions. In the prediction phase, we will again have encoder-decoder architecture. However, we will not train the encoders and decoders since we have trained them already. All we need to do is to change the input parameters.

First, let's discuss how our encoder will look like for making predictions. Take a look at the following script:

```
encoder_model =
Model(encoder_inputs_placeholder,
encoder_states)
```

Here the encoder will be stand alone. Remember, we will not train the encoding model since we have already trained everything (Basically, in training the weights are updated). The weights have already been updated. The only job of the encoder is to take sequence as input and output the final hidden state and cell state of the LSTM.

Next, we need to define the inputs and outputs for our decoder. In the training phase, the inputs to the decoder were the hidden and cell state from the last LSTM unit of the input sequence and the whole target sequence offset by 1. The output was the corresponding target sequence.

In case of predictions, we do not have the whole target sequence since this is what we have to predict. The only thing we have is the hidden and cell state from the input sequence and the start of the word for the target sequence. The output of the decoder will be the corresponding target word and the cell state and the hidden state. This cell state and the hidden state, along with the predicted word will be used as input to the next LSTM unit for the decoder.

Let's create inputs for the decoder LSTM:

```
decoder_state_input_h =
Input(shape=(LSTM_NODES,))
```

```
decoder_state_input_c =
Input(shape=(LSTM_NODES,))
decoder_states_inputs =
[decoder_state_input_h,
decoder_state_input_c]
```

These are the hidden states and the cell state that each decoder LSTM will receive as input.

Next, we need to specify the input shape. Since this time, there will only be a one word as input, the shape of input will be 1. The input is needed to be passed through, embedding layer. The following script does that:

```
decoder_inputs_single =
Input(shape=(1,))
decoder_inputs_single_x =
decoder_embedding(decoder_inputs_s
ingle)
```

Finally, we need to pass the decoder_states_inputs, and decoder_inputs_single_x as the single word to the decoder LSTM, that we trained during the training phase.

```
decoder_outputs, h, c =
decoder_lstm(
   decoder_inputs_single_x,
```

```
initial_state=decoder_states_input
s
)
```

The output of the LSTM will be the decoder_output and the hidden and cell state. The decoder output will be passed through a dense layer to produce the final output as shown in the final script:

```
decoder_states = [h, c]
decoder_outputs =
decoder_dense(decoder_outputs)
```

Notice we stored the hidden and cell state in a variable decoder_states since we want our model to output these values. The sampling model used for prediction can be created using the following script:

```
# The sampling model
# inputs: y(t-1), h(t-1), c(t-1)
# outputs: y(t), h(t), c(t)

decoder_outputs, h, c =
decoder_lstm(
   decoder_inputs_single_x,
```

```
    initial_state=decoder_states_input
s
)
```

In the preprocessing steps, we converted words to integers. The output of our model will be an integer. We need to convert the integer back to the word. Execute the following script to do so:

```
idx2word_input = {v:k for k, v in word2idx_inputs.items()}
idx2word_target = {v:k for k, v in word2idx_outputs.items()}
```

Here, idx2word_input and idw2word target are the dictionaries where the key is the integer and the value is the word that corresponds to the integer. These dictionaries convert words into integers.

Finally, we write a function that takes in the input sequence, and returns back the output sequence. The script for the function looks likes that:

```
def generate_response(input_seq):
    # Encode the input as state vectors.
```

```python
    states_value = encoder_model.predict(input_seq)

    # Generate empty target sequence of length 1.
    target_seq = np.zeros((1, 1))

    # Populate the first character of target sequence with the start character.
    # NOTE: tokenizer lower-cases all words
    target_seq[0, 0] = word2idx_outputs['<sos>']

    # if we get this we break
    eos = word2idx_outputs['<eos>']

    # Create the translation
    output_sentence = []
    for _ in range(max_len_target):
        output_tokens, h, c = decoder_model.predict(
            [target_seq] + states_value
        )
```

```python
    # Get next word
    idx = np.argmax(output_tokens[0, 0, :])

    # End sentence of EOS
    if eos == idx:
        break

    word = ''
    if idx > 0:
        word = idx2word_target[idx]
        output_sentence.append(word)

    # Update the decoder input
    # which is just the word just generated
    target_seq[0, 0] = idx

    # Update states
    states_value = [h, c]
    # states_value = [h] # gru

return ' '.join(output_sentence)
```

The "generate_response" function takes input sequence as the parameter. It first finds the hidden and cell state using the encoder model.

It then creates a new random variable of size 1 x 1 and initializes it with the index of "sos". Remember, the first word of every sentence is "sos". We then find the index of "eos" since we want to end our target sequence generation when "eos" is predicted.

Next, we execute the loop with the number of iterations equal to the maximum target sequence length. During each iteration, we insert our input target plus the previous hidden state values to the decoder_model. In the output, we receive probability distribution of all the possible words along with the hidden state and cell state.

We find the word with the highest probability using the argmax function and store its index. If the index is equal to the index of eos, we exit our loop. Else if the index is greater than zero, we find the corresponding word from the index and append it to the output_sentence list. Finally, we join the individual words in the list and return it.

Step 8: Making Predictions

The final step is to test the chatbots that we just developed. Take a look at the following script:

```
while True:
   # Do some test translations
   i = np.random.choice(len(input_texts))
   input_seq = encoder_inputs[i:i+1]
   translation = generate_response(input_seq)
   print('-')
   print('Input:', input_texts[i])
   print('Response:', translation)

   ans = input("Continue? [Y/n]")
   if ans and ans.lower().startswith('n'):
      break
```

In the script above we create an infinite loop. Inside the loop, we randomly chose an input sequence from the list of input sequences and pass it to the generate_response function which returns the response message. The loop continues until you don't press "n". A snapshot of the output is here attached:

```
Continue? [Y/n]y
.
Input: nice do you use something like that
Response: mess mean ll you

Continue? [Y/n]y
.
Input: what do you think about golang can it replace python
Response: bit

Continue? [Y/n]y
.
Input: never paid attention to it can you recomend some films with that
Response: bit mean you

Continue? [Y/n]
```

The chatbot is not really efficient since I used only 581 samples which is very small compared to the dataset used to develop actual chatbots. However, the concept remains the same; you just need a bigger and better dataset to increase the efficiency of the chatbots.

Conclusion

This chapter marks the end of the book. In this chapter, we saw how we can develop very simple chatbots using LSTM. We saw how sequence to sequence models can be used to model an input sequence to an output sequence. This model can be used to develop any sequence to sequence application.

What to do next?

In this book, we studied some of the most important deep learning concepts. We started with a very basic artificial neural network that performed a simple classification task. We then moved on to a Convolutional neural network to implement image classification. Finally, we saw the usage of LSTM for time series analysis and sequence to sequence problems.

The best way to take most of these concepts that you studied in this book is to find real-world problems and solve them using the different types of neural networks that you studied in this book. For instance, find an image dataset and try to solve classification problems using the convolutional neural network. Another task is to make a translation engine using the sequence to sequence models. Remember, you are going to get better with practice. Start with small tasks and move on to the bigger projects. Happy deep learning.

www.ingramcontent.com/pod-product-compliance
Lightning Source LLC
Chambersburg PA
CBHW020651220526
45464CB00001B/382